T0189732

Studies in Computational Intelligence

Volume 571

Series editor

Janusz Kacprzyk, Polish Academy of Sciences, Warsaw, Poland
e-mail: kacprzyk@ibspan.waw.pl

About this Series

The series "Studies in Computational Intelligence" (SCI) publishes new developments and advances in the various areas of computational intelligence—quickly and with a high quality. The intent is to cover the theory, applications, and design methods of computational intelligence, as embedded in the fields of engineering, computer science, physics and life sciences, as well as the methodologies behind them. The series contains monographs, lecture notes and edited volumes in computational intelligence spanning the areas of neural networks, connectionist systems, genetic algorithms, evolutionary computation, artificial intelligence, cellular automata, self-organizing systems, soft computing, fuzzy systems, and hybrid intelligent systems. Of particular value to both the contributors and the readership are the short publication timeframe and the world-wide distribution, which enable both wide and rapid dissemination of research output.

More information about this series at http://www.springer.com/series/7092

Bipin Kumar Tripathi

High Dimensional Neurocomputing

Growth, Appraisal and Applications

Springer

Bipin Kumar Tripathi
Computer Science and Engineering
Harcourt Butler Technological Institute
Kanpur
Uttar Pradesh
India

ISSN 1860-949X ISSN 1860-9503 (electronic)
ISBN 978-81-322-2894-3 ISBN 978-81-322-2074-9 (eBook)
DOI 10.1007/978-81-322-2074-9

Springer New Delhi Heidelberg New York Dordrecht London

Printed on acid-free paper

Springer is part of Springer Science+Business Media (www.springer.com)

The book is dedicated to the Almighty God, the creator of beautiful world, for enriching my wisdom and my parents for guiding me in my perseverance and blessing me with achievements.

Preface

The neurocomputing is endowed with an opportunity to realize the innate enigma of artificial intelligence in physical world. The researchers of modern computing science have sought to see the neurocomputing as cutting edge technology in computational intelligence and machine learning. The goal of this book is to bring in the elegant theory of neurocomputing that underlies high-dimensional computing and inspire the readers by presentation that the theory is vibrant. It provides a comprehensive foundation of *High-Dimensional Neurocomputing* and represents technology that is rooted in many disciplines. Most significantly, the book emphasizes an aspect of this field that cannot be neglected, that there is a wide experimental side to the practice of *High-Dimensional Neurocomputing*. I tried to strike a balance between theory and practice. It is a speedily expanding field. The book presents a solitary and coherent picture of how to empathize neural function from single neuron to typical networks. I have preferred to present only that material appropriate in constructing a unified framework. Clearly documented and extensively accepted standards presented in the book play a pivotal role for modern computing system. A distinguishing feature of the book is its contents and style of presentation, which I believe, provide an excellent platform for those who want to take up research career in intelligent optimization techniques and engineers who want to implement neurocomputing.

An artificial neuron is the mathematical model of the biological neuron and approximates its functional capabilities. Ever since we invented the idea of artificial neural network, which can learn and generalize, the study of what can be computed and how it can be done well was launched. In the beginning of the twenty-first century, several scientific communities have converged on a common set of issues surrounding various high-dimensional problems. Rumelhart et al. [1] demonstrated that the computing power of an artificial neural network can be enhanced by increasing the number of layers, and this book exhibits that extending the dimensionality of neuron in neural networks originates the similar or better effect on neurocomputing. Boosting the power by extending the dimensionality of neuron in neural networks has been widely accepted for vital high-dimensional applications. A solution to the high-dimensional problem through high-dimensional neural networks, consisting of neuron accepting high-dimensional vector signals (for example, complex numbers and quaternions, N-dimensional vector) is a new directionality for enhancing the ability of neural networks, therefore, is worth researching.

I assume that the readers of the book have a strong background of artificial neural networks in single dimension. In this book, I want to move you above these rudiments by providing you with the tools and techniques, necessary to design and implement high-dimensional neural networks. If we try to identify representation of high-dimensional data as a single cluster (vector or number) than it will be enduring as preferred impinging signals to neuron and surely the refinement in the concepts of neurocomputing. Number or Vector is one of the most elementary notions not only in mathematics, but also in universal science as well. Mathematicians were always fascinated with the possibility of extending the above notion to numbers of high-dimensionality, which follow straightforward algebraic properties. The solution was found when the multi-component numbers or real-valued vectors, as being structures (clusters) subjected to arithmetic properties, were investigated and later successfully utilized in *High-Dimensional Neurocomputing*. This book, inspired by the high-dimensional applications, has led to investigation of many important intelligent computing methodologies. It is the purpose of the book to organize what is known about them in coherent fashion so that students and practitioners can devise and analyze new tools and paradigms for themselves.

In order to provide a high computational power, many attempts have been made to design neural networks, taking account of task domains. Artificial neural network in complex domain is the first and foremost a synthesis of current ideas in theoretical neurocomputing. Complex-valued neural networks whose parameters (input–output, weights and threshold values) are all complex numbers, are suitable for the two-dimensional problems; Not surprisingly, it has also outperformed even for single-dimensional problems. There is another directionality in making the computing power of an artificial neuron high is to devise new form of higher order computing structures. I have adopted and refined the ideas about higher order neurons, both theoretical and practical insight, into the functioning of neural system. I strongly feel that the emphasis on the design as well as on analysis of higher order neuron models in complex domain is the appropriate way to organize the study of neurocomputing. This book does focus on the design of higher order neuron models as well as principles for modeling diverse neural systems. Attempting to construct a general framework for understanding neurocomputing systems provides a novel way to address the theme of the book.

We are seeing a faster and faster move from rule-based system to methods based neurocomputing techniques that learn automatically from very large corpus of examples. They were widely accepted as a rudimentary attempt to generate a coherent understanding of neurocomputing from the perspective of what has become known as machine learning. The machine learning has seen speedy developments since the evolution of high-dimensional neural computing (HDNC). This book discusses many methods related to high-dimensional machine learning that have their bases in different fields: statistics, pattern recognition, biometrics control, computer vision, robotics, etc. HDNC seeks to unify the many diverse strands of machine learning research and to foster high quality innovative applications.

This book presents a readable and concise material to HDNC that reflects the diverse research strands while providing a unified treatment of the field. The book covers all of the main problem formulations and introduces the most important algorithms and techniques encompassing methods from computer science, neural computation, information theory, and statistics. The prerequisites for the successful study of HDNC are primarily a background in traditional neural networks, linear algebra, hypercomplex numbers, multivariate statistics, and differentiability in high dimensions. This book expands and updates coverage of several areas, particularly computer vision, machine learning, and biometric applications that have advanced rapidly over the last decade. The intended audience is quite broad, but mainly consists of the growing number of engineers working in industry, computer scientist, neuroscientist, physicist, and several neurocomputing community interested in learning more about how their quantitative tools relate to the brain. This book is also intended for senior undergraduate, graduate, and seasoned researchers. The researchers will be benefited by discussing the extra research papers; in such case, I hope that the references at the end will provide readers with additional information. My dream in writing this book will be successful if the readers are benefited from this book.

Kanpur, June 2014 Bipin Kumar Tripathi

Reference

1. Rumelhart, D.E. et al.: Parallel Distributed Processing, vol. 1. MIT Press, Cambridge (1986)

Acknowledgments

I am grateful to many researchers from all over the world who have worked hard in this field. I am obliged to my teachers whose inspirational intelligence, experience, and timely advice have been extremely influential in shaping my outlook on life. I am particularly indebted to my teacher Prof. P.K. Kalra, ex-director IIT Jodhpur, Rajasthan (India) for inculcating values and providing inspiring guidance. I would like to thank Mrs. Akanksha Bipin Tripathi for her wholehearted cooperation and constant encouragement given in this successful endeavor. She shouldered a lot of extra responsibilities during the months this book was being written.

I earnestly show appreciation to Mr. Amritanshu and Mr. Prakhar for moral support and unforgettable assistance toward the documentation related to this book. I would like to thank my past and present Ph.D. students; and members of our Nature-inspired Computational Intelligence Research Group (NCIRG) for taking time out for review and valuable suggestions. It has been a pleasure working with the Springer Press and I specially thank Mr. Aninda Bose and Ms. Kamiya Khatter, for all their help and support. I am very sure that the field of *High-Dimensional Neurocomputing* will not only stand still, but also will get significant maturity after this book. I have great pleasure in writing this book; I anticipate you will enjoy reading it.

Contents

1 Neurocomputing: An Introduction 1
1.1 Evolution of Computational Neuroscience 1
1.1.1 Brain Research Chronology 3
1.1.2 Neurocomputing Research Chronology 5
1.2 High-Dimensional Neurocomputing 8
1.2.1 Literature Survey 9
1.2.2 ANN Versus High-Dimensional Neural Network 11
1.3 Neurocomputing in Machine Learning 13
1.3.1 Biometric Application 14
1.4 Scope and Organization of the Book 15
References 19

2 Neurocomputing with High Dimensional Parameters 21
2.1 Neuro-Computing with Single Dimensional Parameters 21
2.2 Neurocomputing in High Dimension 23
2.2.1 Hypercomplex Number System 24
2.2.2 Neurocomputing with Hypercomplex Numbers 26
2.2.3 Neurocomputing with Vectors 26
2.3 Neurocomputing with Two Dimensional Parameters 27
2.3.1 Properties of Complex Plane 28
2.3.2 Complex Variable Based Neural Networks 31
2.4 Neurocomputing with Three Dimensional Parameters 32
2.4.1 Properties of Vector Space 33
2.4.2 3D Vector Based Neural Networks 33
2.5 Neurocomputing with Four-Dimensional Parameters 34
2.5.1 Properties of Quaternionic Space 34
2.5.2 Quaternionic Activation Function 37
2.5.3 Quaternionic Variable Based Neural Networks 38

2.6 Neurocomputing with N-Dimensional Parameters . . . 38
2.6.1 Properties of Vectors in R^N . . . 39
2.6.2 N-Dimensional Vector Based Neural Networks . . . 39
2.7 Concluding Remarks . . . 40
References . . . 41

3 Neurocomputing in Complex Domain . . . 43
3.1 Complex Domain Neuron . . . 43
3.1.1 Why Complex Domain Neuron . . . 44
3.1.2 Out Performance Over Real Domain Neuron . . . 45
3.2 Activation Functions in Complex Domain . . . 47
3.2.1 Why Vary Activation Functions . . . 48
3.2.2 Properties of Suitable Complex-Valued Activation Function . . . 50
3.2.3 Complex Activation Functions . . . 51
3.3 Error Functions in Complex Domain . . . 57
3.3.1 Why Vary Error Functions . . . 57
3.3.2 Definitions and Plots of Error Functions . . . 58
3.3.3 Properties of Complex Error Functions . . . 66
3.4 Learning in Complex Domain . . . 70
3.4.1 Complex Back-Propagation (*C*BP) Learning . . . 71
3.4.2 Complex Resilient Propagation (*C*RPROP) Learning . . . 72
3.4.3 Improved Complex Resilient Propagation (***C***-***i***RPROP) . . . 73
3.5 Concluding Remarks . . . 76
References . . . 77

4 Higher-Order Computational Model for Novel Neurons . . . 79
4.1 Biological Neuron . . . 79
4.2 Artificial Neuron . . . 80
4.2.1 Higher-Order Aggregation Function . . . 82
4.2.2 Why Higher-Order Neurons . . . 82
4.2.3 A Critical Review of Higher-Order Neuron Models . . . 84
4.3 Novel Higher-Order Neuron Models . . . 85
4.3.1 Artificial Neuron Models . . . 86
4.3.2 Model-1 . . . 87
4.3.3 Learning Rules for Model-1 . . . 88
4.3.4 Model-2 . . . 90
4.3.5 Learning Rules for Model-2 . . . 90
4.3.6 Model-3 . . . 91
4.3.7 Learning Rules for Model-3 . . . 93
4.4 Performance Variation Among Conventional and Higher-Order Neurons . . . 95
4.4.1 Real Domain Problems . . . 95
4.4.2 Complex Domain Problems . . . 96

4.5 Concluding Remarks 101
References 101

5 High-Dimensional Mapping 105
5.1 Mapping Properties of Complex-Valued Neural Networks. 105
5.2 Conformal Mapping on Plane 106
5.2.1 Linear Transformation 108
5.2.2 Bilinear Transformation 112
5.3 Concluding Remarks 115
References 116

6 Neurocomputing in Space 117
6.1 3D Vector-Valued Neuron 117
6.1.1 Learning Rule. 118
6.2 Learning 3D Motion 120
6.3 Point Clouds of Objects in Practical Application 123
6.4 3D Face Recognition 123
6.4.1 Normalization. 124
6.4.2 Recognition 125
6.5 Inferences and Discussion. 128
References 128

7 Machine Recognition in Complex Domain. 131
7.1 State-of-Art: Machine Recognition. 132
7.1.1 Effective Feature Extraction 132
7.1.2 Classifier Design. 134
7.2 Multivariate Statistical Techniques in Real and Complex Domain. 137
7.2.1 Feature Extraction with ***R***PCA 137
7.2.2 Feature Extraction with ***C***PCA 139
7.2.3 Independent Component Analysis 140
7.2.4 Entropy and ICA Gradient Ascent. 142
7.2.5 Feature Extraction with ***R***ICA. 145
7.2.6 Feature Extraction with ***C***ICA. 146
7.3 Human Recognition Systems. 148
7.4 Recognition with Higher-Order Neurons in Complex Domain . . . 149
7.5 Performance Evaluation with Different Benchmark Database. . . . 150
7.5.1 Performance in ORL Face Database 150
7.5.2 Performance in Indian Face Database. 155
7.6 Robustness Toward Partial Occlusion, Blurring, and Noises 158
7.7 Inferences and Discussion. 162
References 163

Acronyms

$\boldsymbol{R}$	Set of real numbers
$\boldsymbol{C}$	Set of complex numbers
$\boldsymbol{H}$	Set of quaternionic numbers or quaternions
j	Imaginary unity
$\Re$	Real part of a complex number
$\Im$	Imaginary part of a complex number
Ω	Aggregation function for a neuron
ANN	Artificial neural network
RVNN	Real-valued neural network
CVNN	Complex-valued neural network
VVNN	Vector-valued neural network
QVNN	Quaternionic-valued neural network
CR	Cauchy–Riemann equations
CRF	Cauchy–Riemann–Fueter equations
η	Learning rate
w_{lm}	Synaptic weight from lth Neuron to mth neuron
***R*BP**	Backpropagation learning algorithm in real domain
***C*BP**	Backpropagation learning algorithm in complex domain
f	Real-valued function
f'	First order derivative of function f
f''	Second order derivative of function f
f_C	Complex-valued function
***C*AF**	Complex activation functions
EF	Error function
$[.]^{\aleph}$	Complex conjugate transposition
***C*XOR**	XOR problem in complex domain
AIC	Akaikes information criteria
$\|z\|$	Modulus of a complex number z
Arg	Argument complex variable
MLP	Multi layer perceptron in real domain

CMLP	Multi layer perceptron in complex domain
MSE	Mean square error
cdf	Cumulative distribution function
pdf	Probability density function
$\langle . \rangle$	Expectation computed as the sample average
RPCA	Principal component analysis in real domain
CPCA	Principal component analysis in complex domain
RICA	Independent component analysis in real domain
CICA	Independent component analysis in complex domain
FAR	False acceptance rate
FRR	False recognition rate
3DV-BP	3D vector version of the backpropagation algorithm
RRPROP	Resilient propagation learning algorithm in real domain
CRPROP	Resilient propagation learning algorithm in complex domain
ACONN	All-classes-in-one-neural network
OCONN	One-class-in-one-neural network
OCON	One-class-in-one-neuron

About the Author

Dr. Bipin Kumar Tripathi completed his Ph.D. in Computational Intelligence from IIT Kanpur, India and M.Tech in Computer Science and Engineering from IIT Delhi, India. Dr. Tripathi is currently serving as an Associate Professor in Department of Computer Science and Engineering of HBTI Kanpur, India. He is also leading the Nature-inspired Computational Intelligence Research Group (NCIRG) at HBTI. His areas of research include high-dimensional neurocomputing, computational neuroscience, intelligent system design, machine learning and computer vision focused on biometrics and 3D Imaging. He has published several research papers in these areas in many peer reviewed journals including IEEE Transaction/Elsevier/Springer and other international conferences. He has also contributed book chapters in different international publications and patent in his area. He is continuously serving as PC for many international conferences and as a reviewer of several international journals.

Chapter 1
Neurocomputing: An Introduction

Abstract Human brain is the gateway to the deepest mystery of modern computing science, which confines all the trump cards in the final frontier of technical and scientific inventions. It is an enigmatic and quaint field of investigation having long past and was always being a remarkable interesting area for researchers. The crucial characteristic of human intelligence is that it is evolving (developing, revealing, and unfolding) through genetically 'wired' rules and experience. Neurocomputing is the branch of science and engineering, which is based on human like intelligent behaviors of machines. It is a vast discipline of research that mainly includes neuroscience, machine learning, searching and knowledge representation. The traditional rule-based learning is now appears to be inadequate for various engineering applications because it is incompetent to serve increasing demand of machine learning when dealing with large amount of data. This opened up new avenues for the non-conventional computation models for such applications. Hence, it gives rise to new area of research, which is named as computational neuroscience. This chapter argues that we need to understand evolution of information processing in the brain and then use these principles when building intelligent machines through high dimensional neurocomputing.

1.1 Evolution of Computational Neuroscience

At first glance, the history of the development of computational neuroscience reveals the fact that the basic idea of developing the architecture in a way as to ape the neuronal arrangement in the brain came from a study of the neuroscience, brain anatomy, and analysis of the micro-structure of it. The distinguishing feature that a neuron cell displayed when impinged with a driving force (voltage) resembled the well-known sigmoid function that in a study down the line became accepted as the function of activation of artificial neuron, Fig. 2.1. The neurocomputing from the incipient stages grew as more facts about the human brain surfaced with investigation in Brain Research (Rose and Bynum 1982). In order to process both data and knowledge, neurocomputing evolves its structure (connections and neurons) and functionality in a continuous, self organized, online, adaptive and interactive way through

B.K. Tripathi, *High Dimensional Neurocomputing*,
Studies in Computational Intelligence 571, DOI 10.1007/978-81-322-2074-9_1

genetically defined information and learning. The results of this research that could be improvised, adapted and embedded into the artificial neural network (ANN) got accrued with time and strengthened its performance, brought some limitations to light (Kolmogorov 1957; Werbos 1974; Wang 1992). Before describing the artificial neurons and their networks, it is necessary to consider the biological origin of the artificial neuron.

The three prominent lobes—the big brain or the cerebrum, the middle brain or cerebellum and the hindbrain or medulla oblongata are clearly visible in scientific study of the structure of the human brain. The medulla synapses with the nerves that extend into the spinal column and divide and subdivide as they spread out into the various parts of the body. The nerve endings are fibrous and end in the sense organs. The set of nerves that carry impulses to the brain from the sense organs are called sensory nerves while the ones that take the signal from the brain to the muscles are called mortar nerves. The nerve endings receive impulses and through the network of nerves and pass them to the brain for interpretation. Studies revealed that different parts of the brain are responsible for different faculties. The brain is made up of billions of cells; a rough estimate puts the cells at a hundred billion. Ten billion of these hundred billion cells are 'Neurons' that are responsible for the different functions of the brain. A typical neuron has the central cell or the cell body, the axon, the dendrons that give rise to dendrites. The fibrous dendrites synapse with the cell body and axon of the neighboring neurons. Typically each neuron has 10,000 neurons in its vicinity to which it connects. The conduction of electrical impulse through the structure just described is the mechanism that results in the interpretation of the signal. Neurons respond to the input signal impinged onto them in a certain fashion demonstrate the distinguished feature of neuron cell.

The development in computational neuroscience may conceptually be summarized into biologically inspired technique that has great abilities of learning, association, and generalization. How the neurons operate in the cell body, how they respond or 'fire' to an input impulse; this has been realized in many researches in biophysics but still far from clear understanding. A typical signal to the neuron cell would be of the order of 40 mV, and the response of the Neuron depends on the threshold voltage it operates with. That is, the Neuron fires if the input voltage is greater than its threshold and does not otherwise. Various investigations revealed that the neurons respond according to their characteristic function, also called 'activation function,' which is a general term to describe how a Neuron responds to a stimulus. It is clear that the threshold function is a particular kind of activation function. The first design of ANN came from the study of neuronal arrangement in the brain. The chronological listing referenced from Clarke and Dewhurst (1972), Rose and Bynum (1982), Finger (1994, 2000a), Gross (1998), Marshall and Magoun (1998) brings out the facts with dates. The choice of aggregation function, activation function and the mechanism by which the update of voltage levels happens inside the brain were borrowed in designing the learning rules for the ANN. The new facts about the brain throw more light on the brain's functioning and certainly add to our understanding of its intricate mechanisms. Moreover, the new insights available in functioning formally deliver different paradigms for enhancing our artificial neuron models and aid in our endeavor of bringing the ANN closer to the real functioning of the brain.

The functionalism in artificial neuron is analogous to the biological neuron for machine learning. Neuron is the basic unit in biological neural network having unicellular structures. A neuron comprises of five basic elements: synapse, dendrite, dendron, axon, and cell body. The brain's nerve cells are known as neurons, which transmit and gather electrochemical signals through synapse that are communicated via a network of millions of nerve fibers called dendrites and axons. Till now, the learning process of human beings through neurons is a mysterious prospect. However, the amount of neuro-transmitters released at pre-synaptic end of neurons and its correspondence with learning process demonstrates the participation of neurons in learning process. But exact performance of neurons and their network along with their association with other computational techniques in learning task is an interesting and open research problem in Neurocomputing. The selection of neurons in high dimension is problem-specific, which greatly influences the system performance.

1.1.1 Brain Research Chronology

A survey into historical literature available brings out the fact that investigations in the human brain started 6000 years ago when Sumerians (4000 B.C.) discovered the euphoric effect of poppy seed. The first ever written document on Nervous System called 'Edwin Smith surgical papyrus' emerged during 1700 B.C. Hippocrates (479–360 B.C.) discussed epilepsy as a mental disorder and stated brain as the seat of intelligence. In 280 B.C., Erasistratos of Chios wrote on the divisions of the brain. The institution for the mentally ill was established in 1410 A.D. in Valencia, Spain. In 1536 A.D., Nicolo Massa described the Cerebrospinal Fluid. Andreas Vesalius dissected the human body and published his findings in 'On the Workings of the Human Body' in 1543 A.D. In 1561 A.D., Gabriele Fallippo published 'Observationes Anatomicae' that described some cranial nerves. The term 'hippocampus' was coined in 1564 A.D. by Aranzi and the word 'pons' was coined in 1573 A.D. by Constanzo Varolio. Varolio was the first to have cut the brain at the base and remove the organ from the skull. The sections of the brain began to be studied but a start in this direction was made by Piccolomini in 1586 A.D. when he distinguished the cortex from the white matter. In 1611 A.D., Lazarus Riverius published a text describing impairments on consciousness. In 1649 A.D., Rene Descartes described Pineal Body as the control center of body and mind. In 1664 A.D., Thomas Willis published 'Cerebri Anatome', in Latin and the English version of the same book was published in 1681 A.D. In 1695 A.D., Humphrey Ridley published 'The Anatomy of the Brain'. The word 'reflex' was coined in 1736 A.D. by Jean Astruc. In 1776 A.D., M.V.G. Malacarne published the first book solely devoted to cerebellum.

The respiratory center in medulla was discovered in 1811 A.D. The functional differences between dorsal and ventral roots of the spinal cord was discussed by Charles Bell in 1811 A.D. and by Francois Magendie in 1821 A.D. Jean-Marie-Pierre

Flourens in 1823 A.D. discovered that the cerebellum regulates the mortar activity. In 1825 A.D., Jean-Baptiste Bouillaud presented cases of loss of speech due to frontal lesions. Robert B. Todd discussed the role of the cerebral cortex in mentation, corpus striatum in movement and mid-brain in emotion. In 1826 A.D., Johannes Muller published the theory of specific nerve energies. The cell theory was proposed in 1838 A.D. by Theodore Schwann. The spinal cord's serial sections were studied in 1842 A.D. by Benedikt Stilling. The word 'hypnosis' was coined in 1843 A.D. by James Braid. The six layered cortex was first illustrated by Robert Remak in 1844 A.D. Hermann von Helmholtz measured the speed of frog nerve impulses in 1849 A.D. The nerve galvanometer was invented in 1850 A.D. by Emil Du Bois-Reymond. In 1852 A.D., Kolliker described how mortar nerves originate from the neurons in the anterior horn of the spinal cord. Louis P. Gratiolet in 1854 A.D. described the convolutions of the cerebral cortex. In 1863 A.D., I.M. Sechenov published results on Reflexes of the Brain. John Hughlings Jackson wrote on loss of speech due to brain injury in 1864 A.D. Otto Friedrich Karl Deiters differentiated dendrites from axons in 1865 A.D. In 1869 A.D., Francis Galton claimed the inheritance of intelligence in his publication 'Hereditary Genius.'

The first text book on nervous system surgery came out in 1870 A.D. by Ernst von Bergmann. Richard Caton recorded the electrical activity of the brain in 1875 A.D. David Ferrier in 1876 A.D. published, 'The Functions of the Brain'. In 1877 A.D. Jean-Martin Charcot published, 'Lectures on the diseases of the Nervous System'. The book 'Unilateral Gunshot Injury to the Spinal Cord' was published by W.R. Gowers in 1878 A.D. Emil Kraepelin coined 'neuroses' and 'phychoses' in 1883 A.D. Sir Victor Horsley published the somatotopic map of the monkey mortar cortex. In 1890 A.D., William Ostwald discovered the membrane theory of nerve conduction. In the same year, William James published 'Principles of Psychology.' The term 'neuron' was coined in 1891 A.D. by Wilhelm von Waldeyer. William His used the term 'hypothalamus' in for the first time in 1895 A.D. Adrenalin was isolated in 1897 A.D. by John Jacob Abel. The word 'synapse' was used in the same year by Charles Scott Sherrington. Acetylsalicylic Acid (Aspirin) was synthesized by Felix Hoffman in 1897 A.D. John Newport Langley coined 'Autonomous Nervous System' in 1898 A.D. Aspirin was commercially introduced in 1899 A.D. by Bayer. Sigmung Freud published 'Interpretation of Dreams' in 1900 A.D. Ivan Pavlov coined 'Conditioned Reflex' in 1903 A.D. Alfred Binet and Theodore Simon gave their first intelligence test in 1905 A.D.

In 1906 A.D., Golgi and Cajal shared Nobel Prize for developing a model to describe the structure of the nervous system. Eugen Blueler coined 'schizophrenia' in 1911 A.D. Robert Barany won the Nobel Prize for explaining the Vestibular apparatus in 1914 A.D. Acetylcholine was isolated in the same year. Cecil Vogt described over 200 cortical areas in 1919 A.D. The Society of Neurological Surgeons was founded in 1920 A.D. in the USA. Hans Berger published the first human EEG in 1929 A.D. The multichannel ink-writing EEG machine was developed by Jan Friedrich Tonnies in 1932 A.D. Henry Hallett Dale and Otto Loewi shared Nobel Prize for their work on chemical transmission between nerves. Joseph Erlanger and Herbert Spencer Gasser shared the Nobel Prize on the functions of a single nerve fiber. A.C.A.F. Egas Moniz won the Nobel Prize for developing Leucotomy to treat psychoses. Walter Rudolf Hess won Nobel Prize for work on 'Interbrain.' In 1953 A.D., Eugene Aserinski and Nathaniel Kleitman described the rapid eye movements (REM) during sleep. John Carew Eccles, Alan Lloyd Hodgkin, and Fielding Huxley shared Nobel Prize for work on the mechanisms of the neuron cell membrane. Ragnar Arthur Granit, Halden Keffer Hartline, and George Wald shared Nobel Prize for work on the mechanisms of vision. Konrad Z. Lorenz, Nicolas Tinbergen, and Karl von Frisch shared Nobel Prize for work on Ethology. Roger Guillemin and Andrew Victor Schally shared Nobel Prize for work on peptides in the brain in 1977 A.D. David Hunter Hubel and Torsten N. Wiesel shared Nobel Prize for explaining the visual system in 1981 A.D. In the same year, Roger Walcott Sperry won a Nobel Prize for explaining the functions of the cerebral hemispheres. In 1982 A.D., Bengt Ingemar Bergstrom, John Robert Vane, and Sune K. Bergstrom won Nobel Prize for the discovery of the Prostaglandins. In 1982 A.D., Stanley Cohen and Rita Levi-Montalcini won Nobel Prize for their work on the control of nerve cell growth. In 1991 A.D., Erwin Neher and Bert Sakmann shared Nobel Prize for their work on the function of single ion channels. In 1994 A.D., Alfred G. Gilman and Martin Rodnell shared Nobel Prize for their discovery of G-Protein coupled receptors and their role in signal transduction. For their discoveries concerning signal transduction in the nervous system Arvid Carlsson, Paul Greengard and Eric Kandel shared Nobel Prize in 2000 A.D.

1.1.2 Neurocomputing Research Chronology

Neurocomputing inspired from biological neurons endeavors to achieve human-like ability of intelligence for learning and generalization. Neurocomputing-based techniques have better ability to deal with nonideal scenarios, therefore, they can provide robust and efficient solutions to such situations. Computing derived from the typical convergence of intelligent techniques focuses natural way of problem solving with primarily inclusion of neural network. The mathematical formulation about the convergence of composition of functions was first time demonstrated by Kolmogorov (1957) who stated that any continuous real valued functions in n—variables defined on $[0, 1]^n$ $(n > 2)$ can be represented in the form of composition of function as shown

$$\psi(x_1, x_2, \ldots x_n) = \sum_{j=1}^{2n+1} f_j \left[\sum_{i=1}^{n} g_{ij}(x_i) \right] \tag{1.1}$$

When the function was proposed, it was not apparent how and where it could be applied which left the investigators of the time baffled. Many results followed the discovery of the ANN that stated how and why the neural network algorithms would converge. For instance, there were results on how to select activation functions to ensure the neural network scheme converged (Hassoun 1995). Some efforts describing how to select an architecture to ensure convergence (Mirchandani and Cao 1989) were also reported. It should be however pointed out that most results in the area of neural networks are existential in nature; for instance, to state convergence obtained by an algorithm the theoretical results assure the existence of weights (Rudin 1976) that can approximate the data with functions of the specific form but how the weights can be obtained is not explicitly mentioned.

Artificial neurons offer learning capability which demonstrates artificial intelligence analogous to the human intelligence. ANN-based techniques are robust especially ill-defined problems and can handle uncertainty with ease. Moreover, these techniques are most suitable and provide effective solutions for various real world problems where conventional methods are hard to apply. The earliest history of ANN is limited to the study of McCulloch and Pitts [1], Hebb [2], and Rosenblatt [3] contributions in neuron designing. The historical notes reveal that the first successful attempt to develop the ANN architecture was done by McCulloch and Pitts (1943), hence from the incipient stages since 1943 with the discovery of the perceptron the theory of neural networks traveled vicissitudes of up and down as decades rolled by. The first successful model of artificial neuron did not appear until the Adaptive LINEar combiner (ADALINE) entered the scene in the 1960s and the Widrow-Hoff learning rule that trained the ADALINE-based networks. The ADALINE or the Adaptive Linear Network is the basic element of the neural network that adds the inputs incident onto it. The linear combination with many ADALINES developed MADALINES. The Widrow-Hoff rule minimized a sum-squared error in a pattern classification problem as it trained the MADALINES-based neural network. Research in the direction did not receive momentum as the computational power was insufficient to support the load due to training algorithms; and as a whole, the research in the area slowed down drastically due to the impediment. The idea of multilayered networks was put forth by Nilsson (1965) but his concept could not receive much attention because of lack of interest of researchers. Minsky and Papert (1969) published a book that further put neural networks in jeopardy as it exposed the potential of ANN as computational tools and questioned on the limitations of the perceptron. Research almost stopped after the publication of the work of Minsky and Papert and not much research happened for about 20 years between 1965 and 1984 as the number of researchers in the area dwindled. But the few who pursued Neural Networks during the period made lasting contributions: mathematical theory of neural networks was developed by Sun-Ichi Amari (1972, 1977); Fukushima (1980) developed the Neocognitron, Associative Memory was developed by Tuevo Kohonen (1977, 1980); Cellular neural network by Yang (1988).

Employing the idea of gradient descent, Werbos (1974) worked out on a consistent method (that worked universally) to obtain the weights much after Kolmogorov published the first theoretical result. Werbos' discovery of the BP appeared an important milestone in the history of ANN, though it was not fully free from bottlenecks. The problem of local minima while training using the BP needs a special mention. Depending on the initial condition the network was set to, BP might steer the neural network into a local minimum at which the training process gets stuck (the weights do not get updated but stay frozen even as the epochs run). To circumvent the problem, some algorithms for minimization used the ideas of random search techniques, instead of using the derivatives (Brent 1973). In addition to this, methods to prune synaptic links that are least sensitive to the training process have been proposed (Karnin 1990). This process improves network generalization by decreasing the number of weights, results in a reduced network complexity, and decreases the needed computation. As applied to differential equations, the networks have been used to study chaos in dynamical systems (Aihara et al. 1990). Complex chaotic neural networks were studied by Hirose (1992). Differential Equations were modeled using neural networks and solutions of equations were studied by inputting chaotic initial conditions to the network. Along the same lines but for small architecture neural networks with delay, Francois and Chauvet (1992) reported dynamics of neural networks classifying the regimes as stable, unstable, or oscillatory. Ishi et al. (1996) has applied chaotic neural networks for information processing. Fletcher and Reeves (1964) and Beale (1972) developed the idea of conjugate gradient algorithm. Further, Battiti (1992) reported an efficient method to compute the conjugate gradient. Charalambous (1992) furthered the step by developing a Conjugate Gradient-based BPA incorporating the efficient methods discovered. The scaled conjugate gradient algorithm was put forth by Moller (1993) in which a scaling based on the position in the weights space was used in conjunction with the conjugate direction-based update of weights. A highly efficient way of training using gradient descent, embedding the good features of the second-order algorithms (that involve computing the Hessian) by using an approximation to the Hessian and bypassing the actual computation of it is the Lavenberg-Marquardt Algorithm put forth by Hagan and Menhaj (1994). The neural network is initialized with a set of weights and the performance of the final Network depends on these initial weights. The effect of adding noise during BP training and the final network's performance were studied by An (1995) while the generalization performance based on the weight initialization was studied by Amir et al. (1997). In the same year, Dai and MacBeth reported their observations on learning parameters and how they influence a BPA based training. In order to accumulate the knowledge (training) in the neuron or neural network the variety of learning processes starting from Hebb (1949), Minsky (1961), Ooyen (1992), Haykin (1994) continuing by Riedmiller (1993), Nitta (1997), Tripathi (2011, 2012) were determined to organize the correction of weights.

The size or structure of the neuron architecture that optimally suits a problem remains an open problem in neural networks to this day [4]. Furthering the effort in this direction, the new types of neuron or neural network architectures with multivariable functions and functionals [Cascade Correlation Networks by Fahlman and Lebiere (1990), Pi-sigma Network by Shin and Ghosh (1991), polynomial neural networks by Chen and Manry (1993), higher-order neuron by Schmidt and Davis (1993), the Sum Of Product Neural Networks by Chun (2000)], Quadratic and Cubic neural units by Gupta (2003), multivaluedneuron by Aizenberg (2007) and other non-conventional neural units by Homma et al. (2009), Triron by Tripathi (2010) are few of them were proposed. The compensatory neuron structure were employed by Kalra in 2000 for control problems of the aerospace industry and to determine satellite orbit motion. The idea of compensatory structure was better established by B.K. Tripathi (among summation, radial basis function, and product) and proposed compensatory nonconventional neuron models in real and complex domain in 2010 and obtained an accelerated learning. The unsupervised learning and image processing techniques were further coupled with neural network to detect and circumvent image processing and computer vision applications to test the effectiveness of the neurocomputing. In 2011, Tripathi used family of quasi-arithmetic means that covers the entire interval of averaging operations between minima and maxima, for an aggregation function and demonstrated root power-mean neuron (RPN 2011), which is naturally general and includes various existing artificial neurons as its special cases. The successive chapters of this book consider these issues as basis for better explanation of high dimensional neurocomputing with higher-order neural processing.

In due course of time, the idea of changing the error function and activation function also evolved for better optimization in neurocomputing. Werbos and Titus (1978), Gill and Wright (1981) attempted different error functions in an optimization scheme. Later in 1983, Rey stated that by varying the error function in an optimization scheme, the result could be improved substantially. The statement was backed by demonstration (Rey 1983) that an Absolute Error Function-based optimization solved a curve-fitting problem more efficiently than the standard Quadratic Error Function-based optimization. Fernandez (1991) implemented some new Error Functions that were designed to counter the ill-effects local minima by weighting the errors according to their magnitudes. Matsuoka (1991) reported BPA based on logarithmic Error Function and elimination of local minima. Ooyen and Nienhaus (1992) used an entropy-type Error Function and showed that it performs better than the Quadratic Error Function-based BP for function approximation problems. Similarily, wide variety of activation functions have also been considered for different optimization schemes. Huang and Babri (1998) reported a result that spelt out an upper bound to the number of hidden neurons for neural networks with arbitrary nonlinear activation functions. Guarneri and Piazza (1999) reported an adaptive spline-type activation function for standard neural network-based training. The varietation in activation functions is more remarkable when dealing with high-dimensional neurocomputing. The analyticity (diiferentiability) and boundedness of the functions are imortant issues in high dimensions. A wide variety of activation functions have been investigated including the analytic function [5], local analytic function [6, 7], split-type function [8, 9], phase-preserving function [10, 11] and circular-type function [12] for neural networks dealing with high-dimensional parameters. Chapter 3 present and critically analyze the error functions and activation functions for high-dimensional neurocomputing.

1.2 High-Dimensional Neurocomputing

The evolution of neurocomputing has envisaged an advanced and a mechanized world, where human life is made better by developments in biologically inspired techniques. These technologies make an important bridge to computer science and other discipline that study information processing even may be in high dimensions. An improved understanding of how a single neuron operates has lead to an concrete development of high-dimensional neurocomputing techniques. High-dimensional neural networks accept and represent information in different components of high dimensions as a single entity (cluster) thus allowing the processing of magnitude and phase of a point simultaneously. Moreover, extensive studies carried out during the past few years have revealed that high-dimensional neurocomputing paradigms enjoy numerous practical advantages over conventional neurocomputing (deal with real-valued data or in single dimension) applied for high-dimensional information. They have proved to be powerful mathematical instrument for modeling typical systems. Recently, there has been an increasing interest in high-dimensional neuro-computing. It provides an easy, fast, and specific implementation of operations through high-dimensional neural networks.

1.2.1 Literature Survey

Neuro-computing comprises of biological-inspired computing methodologies and techniques are capable enough to address complex problems of the real world for which conventional methodologies and approaches are ineffective or computationally intensive. The history of high-dimensional neuro-computing starts with the development of CVNN which can be traced to the ideas presented by N. Aizenberg in 1971 in the Soviet Union [13]. This direction was related to neurons with phase dependent functions. These ideas were later developed by I Aizenberg in the form of multivalued neurons and universal binary neurons [14, 15]. The research in the area took a different turn in the early 1990s with the publication of Back-Propagation Algorithm in complex domain (*C*BP). But the complex version of the BP made its first appearance when Widrow, McCool and Ball (1975) announced their Complex Least Mean Squares (LMS) algorithm. Kim and Guest (1990) published a complex-valued learning algorithm for signal processing applications. The necessity came in the form of capturing the phase information in signal processing applications where complex numbers naturally enter the study and must be retained all through the problem as they should be later interpreted. Lueng and Haykin (1991) published the *C*BP in which the activation used in complex domain was an straight forward extension of the sigmoid function. Georgiou and Koutsougeras (1992) published another version of the *C*BP incorporating a different activation function. The dynamics of complex-valued networks was studied by Hirose (1992), which was later applied to the problem of reconstructing vectors lying on the unit circle. Benvenuoto and Piazza (1992) developed a variant of the *C*BP by extending a real activation function in complex domain differently. A complex-valued recurrent neural network was proposed by Wang (1992) that solved complex valued linear equations. Deville (1993) implemented a complex activation function for digital VLSI neural networks that required lesser hardware than the conventional real neural network would need. An extensive study of the *C*BP was reported by T Nitta (1997) in which a learning algorithm along with complexity analysis was presented. He also pointed out certain problems that the standard BP fails but the *C*BP manages to solve. He and later Tripathi (2009) prefered to incorporate nonanalytic (so called "split") but bounded activation function and demonstrated the reasons of his preference. Meanwhile, T Adeli et al. (2003) fully complex-valued neural network with different analytic activation functions with the arguments to overcome the issues related the unboundedness of function [5]. The set of problems to which the *C*BP can be successfully applied is a research theme in the light of the problems that the *C*BP solves but the BP fails to do so. From 2009 to 2012, Tripathi did an exhaustive study over related researches and compiled this diverse field into a comprehensive and novel complex-valued neurocomputing. All these works were endeavors toward the development of high-dimensional neurocomputing along with its performance evaluations and wide applicability.

Various researchers have independently proposed extension of real-valued neuron (one dimension) to higher dimensions [10, 16–18]. Most of them have followed natural extension of number field like real number (one dimension), complex number (two dimension), 3D real-valued vectors (three dimension), quaternion (four dimension), etc., for representation of higher dimension neurons. Complex-valued neural network has received much attention from researchers in the recent past. It can directly operate with 2-D information. Thus, it is more significant in problems where we wish to precisely learn and analyze signal amplitude and phase simultaneously. The researches in biophysics highlight the fact that the action potential in the human brain may have different pulse patterns and the distance between pulses may be different. This justifies the introduction of complex numbers representing phase and amplitude into neural networks. CVNN provide efficient solution for complex-valued problems. Tohru Nitta [8] and Tripathi [11] had shown that the number of learning parameters needed with CVNN for a complex-valued problem is approximately half the equivalent RVNN, even when each complex-valued parameter is counted as two. They also shown that time complexity per learning cycle is same in both networks, but the number of learning cycles in CVNN are much less than the equivalent RVNN. CVNN is more general in the sense that it can provide efficient solution for functions in single dimension as well as on a plane (two dimension). The performance of CVNN has also scored over RVNN even on real-valued problems. For example in [19, 20], a single layer CVNN has successfully solved the XOR problem, which cannot be solved by a single layer RVNN. Moreover, an efficient solution for many real-valued problems has been achieved with CVNN in few recent publications [14, 21, 22]. They are universal approximators. For the problem of same complexity, they also require smaller network topology (hence lesser number of learning parameters) and lesser training time to yield better and more accurate results in comparison to equivalent RVNN. Its 2D structure of error propagation reduces the problem of saturation in learning and offers faster convergence.

The application field of CVNN is very wide. It is not easy to imagine areas dealing with 2-D parameters without the realm of complex numbers. Applications in new technologies such as robotics, signal processing, intelligent systems, communication, space and ocean technology, medical instrumentation as well as in older technologies, namely control and prediction problems are creating a wide spectrum of examples for CVNN, in which nonlinearities, uncertainty, and complexities play a major role. A number of CVNN hardware have also been developed using electromagnetic and light waves. Further, quantum neural network has also became an emerging field of investigation in recent researches.

Though complex-values can treat two-dimensional data elements as a single entity, what we should treat data with more than two-dimension in artificial neural networks? Although this problem can of course be solved by applying several real-valued or complex-valued neurons, it would be useful to introduce a number system with higher dimensions, the so-called hypercomplex number systems. It is imperative to look for higher dimensional neuron model that can directly process the high-dimensional information. It will serve as a building block for a powerful ANN with fewer neurons. In the beginning of twenty-first century, researches devoted thier investigation for neural network with 3D vectors and in the quaternionic domain. They proposed learning machine consisting of multilayer network of 3D vector-valued and quaternion neurons. The equivalent error back-propagation training algorithm for 3D vector-valued and quaternionic neural network is the natural extension of complex-valued backpropagation (BP) algorithm and has natural ability to learn high dimension motion compactly and efficiently as $\mathbb{C}$BP learns 2D motion.

1.2.2 ANN Versus High-Dimensional Neural Network

An ANN is a biologically inspired technique that has great abilities of learning, association, and generalization [23, 24]. It relies largely on parallel processing and is used in decision making based on incomplete data. Extensive studies carried out during the past several years have revealed that neural networks enjoy numerous practical advantages over conventional methods. They have proved to be powerful mathematical instrument for modeling complex systems. Traditional neural network's parameters are usually real numbers, which deal with real-valued data or applicabe for single dimension problems. Recently, there has been an increasing interest in ANN with high-dimensional parameters. They provide an easy implementation of operations in high dimensions. Complex-valued, vector-valued, and Quaternion-valued neural networks (CVNN, VVNN, and QVNN) have been introduced to characterize high-dimensional information effectively. Their functionality depends on mathematical theories that define nonlinear high dimension functions. The developments in this area presented the second generation of neural network. Processing multidimensional data is an interestingly important problem for artificial neural networks. A single conventional neuron can take only one real value as its input, therefore their network should be configured incorporating several neurons and typical mesh connections for processing multidimensional data. This type of configuration is sometimes unnatural and computation itensive in applications of artificial neural networks. It is also not possible to learn and generalize phase among different components simultaneously with the magnitude in high-dimensional data through conventional real-valued neural networks.

The complex number systems have been utilized to represent two-dimensional data elements as a single entity in CVNN. The application of complex numbers to neural networks have been extensively investigated [10, 11, 25, 26]. As can be noted, the CVNN looks exactly like the ANN as the neuron, how it operates, the architecture are all similar. The difference lies is the fact that the weights, input–output, and bias are complex numbers in the CVNN unlike ANN where they were real numbers. The activation functions in the CVNN are complex valued in opposition with the standard ANN where the functions were real valued. It is obvious that the theory of complex variables and complex-valued functions must be applied for studying the behavior of the CVNN. The activation function too was an extension of what existed as the function of activation in the ANN but needed an improvisation to suit to the complex variable-based ambience. In the process, new constraints surfaced (in the form of Liouville Theorem) that had to be cleared for which a different search had to be carried out. Importantly, CVNN have the potential power in analyzing typical functions on a plane. They have also presented improved results even in case of real-valued problems. This is because they are more efficient, fault-tolerant, less sensitive to noise, and better at mimicing human-like characteristics (learning and generalization) in single- and two-dimensional situations. The development of the CVNN came about as an extension of the ANN and not as a prototype of the neuronal arrangement in the brain. In essence, the CVNN is an extension of the ANN but does not draw anymore from the actual neurons and their arrangement in the brain. The complex-valued signals flowing through a complex domain network are the unit of learning, which enable to learn 2-D motion of signals. In contrast, a neural network in a real domain administers 1-D motion of signals. Thus a neural network extended to complex domain has phase preserving nature during learning, while an equivalent real domain network cannot. It is worth to mention here that high-dimensional neural network does not learn only the magnitude along different dimensions but also phase along different directions. Therefore, conformal mapping on plane preserves the angles between oriented curves and the phase of each point on the curve is also maintained during transformation.

Quaternion is a four-dimensional hypercomplex number system introduced by Hamilton [27]. This number system has been extensively used in several fields, such as modern mathematics, physics, control of satellites, computer graphics, etc. [7]. Applying quaternions to the field of neural networks has been recently explored in an effort to naturally adapt and represent high-dimensional information by a quaternionic neuron, rather than complex-valued or real-valued neurons. Thus, there has been a growing number of studies concerning the use of quaternions in neural networks. In principal the neuron in quaternionic domains, which is a four-dimensional hypercomplex number system, can be decomposed and represented by two complex numbers with two linearly independent bases.

Number is one of the most elementary notion not only in conventional computing but also in high-dimensional computing. We are very much fascinated with the possibility of extending the notion of numbers to the high dimensions, which follow straightforward algebraic properties. These multy-component numbers were successfully utilized in high-dimensional neural networks. Unfortunately, the basic algebraic properties can only be satisfied by multy-component numbers only in dimensions 1, 2, 4, and 8 that is in real numbers, complex numbers, quaternions and octonions. Such generalization is not possible for three-dimensional numbers [27, 28] and in other higher dimensions. Here is another way for extending the dimensionality of neural networks to dimensions is through the use of real-valued vectors which originates the similar effects on neural networks. This notion originates an N-dimensional vector neuron, which can deal with N impinging signals as one cluster (single entity). It conveys the potent computational power of N-dimensional vector neurons which corresponds to the ability of the complex-valued neuron in two dimensions. The decision boundary of a N-dimensional vector neuron consists of N hyperplanes which intersect orthogonally each other and divides a decision region into N equal sections, as in the case of a complex-valued neuron. This direction of making the dimensionality of artificial neuron high, taking account of task domain, will offer a high computational power in high-dimensional neural networks.

The functions of activation for high-dimensional neurons have a very promenent role in structure, characterestic and performance of corrosponding high-dimensional neural networks. A wide variety of activation functions for high-dimension neurons have been investigated including the phase-preserving function, circular-type function, split-type function, locally analytic function, analytic function etc. [5, 6, 8, 12, 29]. Some time these functions are contingent upon the problem and some times depends on the architecture of artificial neuron. The quality of the functions also put their impact on the nature of derived learning algorithms. Application of the presented network to engineering problems is also challenging. The processing of three or four dimensional vector data, such as color/multispectral image processing, predictions for three-dimensional protein structures, and controls of motion in three-dimensional space, will be the candidates from now on. Successive chapters take care of very specific example problems of different domains of application to present the motivation and theme of the book.

1.3 Neurocomputing in Machine Learning

The researchers in the artificial intelligence and related areas have sought to see the brain as simply a glorified computing machine. The highly distributed cooperative computation deepens our understanding of the human brain and catalyzes for the development of computing machinery. Using metaphors to describe the brain, distributed computing is successfully tried in neurocomputing that underlie intelligence in the machine. The machine learning is mainly concerned with application of artificial intelligence over large data base. With innovations in computer technology, the neurocomputing techniques have became very prominent for machine learning applications. The machine learning also assists us in designing intelligent systems and to optimize their performance using example data or past experience. An intelligent system should have ability to learn and adapt in a varying environment, then the system designer need not to foresee and impart answers for all possible solutions.

Nowadays, machine learning uses the theory of statistics and neurocomputing in building intelligent systems, because the core task is making inference from sample and mathematical models.

Modern computers have revolutionized our globe. They have utterly transformed the track of our daily life, the way we do science, the way that agriculture and business is continued, the way we entertain ourselves and the way we shield out security. Machine learning is programming the computers to optimize performance criterion using example data and artificial intelligence techniques. This requires to code the given learning rules in a appropriate programming language, devise suitable datasets and write testing/generalization program which output and analyze the result. It is better to adapt the system for general circumstances, rather than explicitly writing a different program for each special circumstance. The responsibility of computer is twofold. First is iteratively train the system using training data to optimize the parameters of the computing model for solving optimization problem. A learning routine is able to adapt to the basic characteristic by monitoring the example dataset. Second is to store optimized parameters and process the huge amount of data for output and analysis. The efficiency of computing system in terms of time and space may be equally important as its predictive accuracy in intelligent system design. An example is an intelligent user interface that can adapt to the biometrics of its user—namely, his or her accent, facial features, iris, oculomotion characteristics, and so forth [30]. The promising capabilities of high dimensional neurocomputing in the adaptability, context-sensitive nature, error tolerance, large memory capacity, reduction of the computational efforts, and real-time capability of information processing in single and high dimensions suggests an alternative architecture to emulate for machine learning.

1.3.1 Biometric Application

Biometric applications refers to the identification or recognition of individuals based on unique physiological or behavioral characteristics or traits. The relevance of biometrics with computer and information science is much broader in terms of capability of the fast computing machines for huge information processing even better than the human intelligence. In recent past, several techniques and biometric traits are developed for human recognition. The basic concept is to use special characteristics of an individual and identify him based on this special property. One can divide special characteristics into two categories, viz physiological (face, iris, finger print, periocular, palm print, hand geometry) and behavioral (voice, eye-movement, gait) characteristics. In biometrics, the neurocomputing-based techniques are interestingly used because of better performance over the conventional statistical techniques. The key reasons that make neurocomputing competitive for biometric applications may be summarized as follows:

- Neurocomputing-based practices are essentially characterized by imitating biological functions, which exhibits natural candidate technology in biometric research.
- Biometric recognition can be customarily dignified as typical nonlinear problem. The neurocomputing-based techniques have appeared very successful in solving these typical problem.
- Since neurocomputing-based techniques have better ability to deal with nonideal scenarios, therefore, they can provide robust and efficient solutions in the varied environmental situations which include variations in illumination, noise, pose, and partial occlusion.
- The neurocomputing-based techniques are efficiently adaptive. This capacity allows the development of real time biometric system capable of online learning and able to conveniently adjust if new subject is added in the biometric data base.

Biometric recognition systems have shown excellent performance in the field of secured access control, forensic science and surveillance using face, palm, fingerprint, iris, periocular and oculo-movement scan path signals and other biometric traits for human recognition [30]. The fundamental requirement of any biometric recognition system is a human trait having several desirable features like universality, distinctiveness, permanence, collectable, performance, circumvention, and acceptability. However, a human characteristic possessing all these features has not yet been identified. As a result, none of the single biometric trait can provide perfect recognition. It is also difficult to achieve very high recognition rates using single trait due to problems like noisy sensor data and nonuniversality or lack of distinctiveness of the chosen biometric trait. Therefore, the performance of a biometric system may be improved by utilizing a number of different biometric identifiers. The result of combined effects will be more robust to noise; and minimize the problem of nonuniversality and lack of distinctiveness. The face recognition, one of the most important biometric traits due to its non intrusive nature, has been considered in this book for illustration of computing capabilities of high-dimensional neurocomputing. Results demonstrated in Chap. 7 on face biometric trait is very competitive with respect to related existing techniques.

1.4 Scope and Organization of the Book

The second generation neurocomputing may conceptually be defined as the era of high-dimensional neural networks, higher-order neurons and fast supervised and unsupervised learning algorithms. These advances have prompted active research activities in neurocomputing where it was supposed to be saturation in the first generation. The artificial neural networks with high-dimensional parameters are developed to explore natural processing of high-dimensional data, considering them either in the form of number or vector. They are characterized on the basis of high-dimensional information, which flows through the network. They lead to adaptive systems which can be simulated on digital computer. In fact, this direction has provoked an effective development in machine intelligence, bioinformatics, computer vision and other computer science and engineering applications. An artificial neuron is the simplified model of a biological neuron which can approximate its functional capabilities. But, for the time being, it is far from clear how much of this simplicity is justified, since at present we have only a poor understanding of neuronal functions in biological neuron. Conventional neurons, based on radial basis function (Rbf) or summation aggregation function, were thoroughly used in first generation ANN. However, networks based on these neurons take a large number of neurons which increases the complexity of the topology, training time and memory requirement in experiments. The problem is circumvented by higher-order neurons in second generation. Their networks have shown improved results with fewer neurons. However, they suffer from a typical curse of dimensionality due to a combinatorial explosion of terms [31–33]. Therefore, it is desirable to investigate some potential neuron models which capture nonlinear correlation among input components but are free from this problem. This book also focuses on the design and assessment of such higher-order neurons. Supervised learning schemes in high dimensions are discussed. Unsupervised learning in a complex domain has been presented for extraction of lower dimensional features with significant discriminating power. Many real applications in the areas of adaptive computing involve signals that are inherently high dimensional. The physical characteristics of these signals and their nonlinear transformations can be approximated efficiently if they are represented and operated as a cluster (single entity) of component signals. The development of high-dimensional neural networks to preserve and process these signals in high dimensions itself is gaining more attention.

The theories and practices presented in this book is an attempt to bridge the gaps among the prominent concepts in second generation neurocomputing. Though theory is getting maturity but the applications are just beginning to be understood, which is clear from the work available in the area. Most practical problems that come from various fields (Robotics, Medicine, Industry, Military, Aviation, and so on) that involve modeling with high-dimensional neural networks. Their applications will become clear once they are applied to some standard problems. Therefore, major issues raised are addressed in the book by applying high-dimensional neural network to the problems of classification, approximation, function mapping, and pattern recognition. The theory and application of complex variable-based neural networks, the most basic case of high dimensional neurocomputing, have been given rigorous attention from the viewpoint of the second generation neurocomputing. The update rule is exactly same as the one used while running the conventional error BP algorithm to train an ANN. However, it must be noted that the complex number comes with the phase information embedded into it. This amounts to saying that the information that would have separately been input to the ANN while training (as is usually done while training ANNs) gets coupled resulting in a decrease of the number of inputs by as much as half (as two real numbers make one complex number and phase information is embeded in it) at the same time preserve the information in the form of phase. There exist additional constraints of analyticity of functions in complex variable setup. It is hence not clear how the new learning algorithm would perform, working with the constraints of analyticity and/or boundedness, when applied to various problems. In most problems of practical interest that involve an optimization process, a quadratic error function is chosen and subject to an optimization. It was pointed out in the literature (Werbos and Titus 1978; Gill and Wright 1981; Fernandez 1991) that employing a different error function can improve the performance of an optimization scheme. Moreover in the m-Estimators approach to data analysis (Rey 1983), a number of functions that can effectively serve as error functions have been listed. It was shown that the new error functions have the ability to suppress the ill-effects of the outliers and exhibit a robust performance to noise and outperform the standard quadratic functions when applied to optimization problems involving data with a scatter of outliers. Hence the question of how the error BP algorithm would perform when the error function is varied, immediately comes up. Moreover, recently it has been shown that the complex-valued neural networks help to solve real-valued problems more efficiently than their real-valued counter-parts [21, 26]. Since then, several complex-valued neural networks have been developed to solve real-valued problems [9, 14, 34, 35]. Further, the book extends the theory and practice for vector-valued neural networks for those who are interested in learning more about how to best characterize the systems and explore experimentally in three-dimensional neurocomputing.

The scientific community believe that for artificial intelligence to be a reality, that is system to be as intelligent as us, we require relatively simple computational mechanism, then only one can simulate the real aspects of human intelligence in broad way and can totally achieve it. Unfortunately, there is lack of compiled literature aimed at providing clear and general methods for high dimensional neurocomputing tools at the systems level. More practically talking, we also desire to promote the current trend for experimentalists to take sincerely the insights gained from using high-dimensional computing. The explicit methodology provided and many examples presented in the book are intended to show precisely how high-dimensional neurocomputing can be used to build the kind of mechanism that scientific community and other readers can exploit.

This book is great for students and professionals as well as would be a nice supplement for researchers taking career in a new generation of neural networks. Again, I cannot believe how readable the book is. The prospective readers who are not familiar with this wonderful philosophy would certainly enjoy the book. Each chapter start with description of strategy in general terms which outline the form that computation will take in detail inside the different sections. The major goal of the organization of chapter is to emphasize the arts of synthesis and analysis of neurocomputing. The respective chapters are organized as follows:

This chapter introduces the reader with the evolution of computational neuroscience and chronological developments in the history of neurocomputing. This also presents the supremacy of high-dimensional neurocomputing in various recent works which continue to be a compelling reference work for advances in machine learning and intelligent system design.

Chapter 2 is devoted for those interested in knowledge more about how their quantitative representation of signals flowing through ANN relate to the neurocomputing and discuss how to understand the high-dimensional neurocomputing using familiar techniques of vector algebra and hypercomplex number system. Now, if readers are not familiar with the generalization of real numbers such as complex numbers and quaternions, this is an excellent introduction to them. The comprehensive extension in the dimensionality of neurocomputing originates a sense of which neural systems are appropriate targets for particular kinds of computational modeling; and how to go about modeling such systems. This concept is important for those readers who are less accustomed with the wonderful high-dimensional neurocomputing in general would certainly enjoy the later parts of this book.

Chapter 3 sets the stage for the in-depth coverage of neurocomputing in complex domain. Another work focusing on various types of neural activation functions and their differentiability can be found in the chapter. Next is a discussion of different types of activation functions followed by a presentation of error functions. The conventional error BP algorithm minimizes a quadratic error function by steering the weights along the direction of negative gradient (using the update rule), the chapter points out alternative error functions that can effectively better the performance of a neural network. Next, more advanced training methods are given for error BP learning.

Chapter 4 begins with a discussion of basic neuron models that are the building blocks of neural system. From recent publications, chapter mentioned that the neuron models with nonlinear input aggregation have better computational and learning properties than conventional neurons. The use of nonconventional neural units (higher-order neuron) appears to take over popularity over conventional in most recent publications. They offer adjustable strong nonlinear input–output mapping without local minima issue due to nonlinearity of the neural architecture itself. This chapter introduces two compensatory higher-order neural units; and one generalized higher-order neural units as the various existing standard models are its special case. The theoretical derivations of learning rules are supported by examples that demonstrate the superiority of presented approach.

Chapter 5 presents functional mapping properties of high-dimensional neurons to demonstrate their phase approximation capabilities. For phase approximation, one does not need to use an error function which simultaneously minimizes both the magnitude and phase errors. Because of inbuilt nature of complex numbers (real and imaginary parts along with embedded phase information), which flow through complex-valued neural network. Thus, the learning algorithm achieves convergence not only with respect of magnitude but also with respect to phase. Therefore, during functional mapping (transformation) the phase (angle) property of each point in a geometric structure remains preserved not only in magnitude but also in sense. The illustrative examples in the book demonstrate the phase preserving property through variety of mapping problems on plane (conformal mapping). The concept presented in this chapter has wide spectrum of applications in science and engineering, which further need investigation.

Chapter 6 formulates a model neuron that can deal with three-dimensional signals as one cluster, called 3D real-valued vector neuron. The basic learning rules for training sophisticated network of three-dimensional vector neuron are covered. The chapter concludes with the performance analysis over selected class of problems. The functional mapping properties of two-dimensional neurons are further extended in this chapter for three-dimensional neuron. The three-dimensional face recognition problem discussed in the chapter opens a path for more biometric applications.

Chapter 7 introduces an approach for evaluating, monitoring, and maintaining the stability of adaptive learning machine for prospective applications. The approach allows us to evaluate the capability of supervised and unsupervised learning of neural units in complex domain that are a fundamental class of high-dimensional neurocomputing. The chapter starts with an overview of selected data preprocessing and feature extraction. It presents PCA and ICA algorithm in real and complex domain for feature extraction along with optimal neural recognizer (OCON: One-Class-in-One-Neuron) for statistical analysis of data. The readers will be most interested in how this book brings about the computation capability of single neuron and how one can apply considered techniques for generating the large-scale realistic simulation. The machine learning applications have been analyzed by typical real life problem of biometric applications. The illustrative examples in the book demonstrate the improvement in learning capability in terms of speed and performance with less number of neurons and learning parameters as compared to the standard neural networks, hence reduces the overall burden of the network.

References

1. McCulloch, W.S., Pitts, W.: A logical calculation of the ideas immanent in nervous activity. Bull. Math. Biophys. **5**, 115–133 (1943)
2. Hebb, D.O.: The Organization of Behavior. Wiley, New York (1949)
3. Rosenblatt, F.: The percepton : a probabilistic model for information storage and organization in the brain. Psychol. Rev. **65**, 231–237 (1958)
4. Tripathi, B.: Novel neuron models in complex domain and applications, PhD Thesis (Thesis defended on 21 April 2010), Indian Intitute of Technology, Kanpur (2010)
5. Kim, T., Adali, T.: Approximation by fully complex multilayer perceptrons. Neural Comput. **15**, 1641–1666 (2003)
6. Isokawa, T., Nishimura, H., Matsui, N.: Quaternionic multilayer perceptron with local analyticity. Information, **3**, 756–770 (2012)
7. Matsui, N., Isokawa, T., Kusamichi, H., Peper, F., Nishimura, H.: Quaternion neural network with geometrical operators. J. Intell. Fuzzy Syst. **15**, 149–164 (2004)
8. Nitta, T.: An extension of the back-propagation algorithm to complex numbers. Neural Netw. **10**(8), 1391–1415 (1997)
9. Tripathi, B.K., Kalra, P.K.: Complex generalized-mean neuron model and its applications. Appl. Soft Comput. **11**(01), 768–777 (2011)
10. Hirose, A.: Complex-Valued Neural Networks. Springer, New York (2006)
11. Tripathi, B.K., Kalra, P.K.: The novel aggregation function based neuron models in complex domain. Soft Comput. **14**(10), 1069–1081 (2010)
12. Georgiou, G.M., Koutsougeras, C.: Complex domain backpropagation. In: IEEE transaction on circuits and systems-II: Analog and Digital Signal Processing, vol. 39, no. 5, pp. 330–334, May 1992
13. Aizenberg, N.N., Yu, I.L., Pospelov, D.A.: About one generalization of the threshold function. Doklady Akademii Nauk SSSR (The Reports of the Academy of Sciences of the USSR) **196**(6), 1287–1290 (1971) (in Russian)

14. Aizenberg, I., Moraga, C.: Multilayer feedforward neural network based on multi-valued neurons (MLMVN) and a back-propagation learning algorithm. Soft Comput. **11**(2), 169–183 (2007)
15. Aizenberg, I., Aizenberg, N., Vandewalle, J.: Multi-valued and Universal Binary Neurons: Theory, Learning, Applications, ISBN 0-7923-7824-5. Kluwer Academic Publishers, London (2000)
16. Tripathi, B.K., Kalra, P.K.: On the learning machine for three dimensional mapping. Neural Comput. Appl. **20**(01), 105–111 (2011)
17. Nitta, T.: A quaternary version of the backpropagation algorithm. Proc. IEEE Int. Conf. Neural Netw. **5**, 2753–2756 (1995)
18. Nitta, T.: N-dimensional Vector Neuron. IJCAI workshop, Hyderabad (2007)
19. Nitta, T.: Orthogonality of decision boundaries in complex-valued neural networks. Neural Comput. **16**(1), 73–97 (2004)
20. Nitta, T.: Solving the XOR problem and the detection of symmetry using a single complex-valued neuron. Neural Netw. **16**, 1101–1105 (2003)
21. Amin, M.F., Murase, K.: Single-layered complex-valued neural network for real-valued classification problems. Neurocomputing **72**(4–6), 945–955 (2009)
22. Chen, X., Tang, Z., Li, S.: A modified error function for the complex-value back propagation neural networks. Neural Inf. Process. **8**(1), (2005)
23. Zurada, J.M.: Introduction to Artificial Neural Systems. Jaicob publishing house, India (2002)
24. Gupta, M.M., Homma, N.: Static and dynamic neural networks. Fundamentals to Advanced Theory, Wiley (2003)
25. Mandic, D., Goh, V.S.L.: Complex Valued Nonlinear Adaptive Filters: Noncircularity, Widely Linear and Neural Models. Wiley, Hoboken (2009)
26. Tripathi, B.K., Kalra, P.K.: On efficient learning machine with root power mean neuron in complex domain. IEEE Trans. Neural Netw. **22**(05), 727–738, ISSN: 1045–9227 (2011)
27. Hamilton, W.R.: Lectures on Quaternions. Hodges and Smith, Ireland (1853)
28. Schwartz, C.: Calculus with a quaternionic variable. J. Math. Phys **50**, 013523:1–013523:11 (2009)
29. Piazza, F., Benvenuto, N.: On the complex backpropagation algorithm. IEEE Trans. on Signal Process. **40**(4), 967–969 (1992)
30. Srivastava, V.: Computational Intelligence Techniques: Development, Evaluation and Applications. PhD Thesis (Thesis defended on 26 April 2014), Uttar Pradesh Technical University, Lucknow (2014)
31. Taylor, J.G., Commbes, S.: Learning higher order correlations. Neural Netw. **6**, 423–428 (1993)
32. Chen, M.S., Manry, M.T.: Conventional modeling of the multilayer perceptron using polynomial basis functions. IEEE Trans. Neural Netw. **4**, 164–166 (1993)
33. Kosmatopoulos, E., Polycarpou, M., Christodoulou, M., Ioannou, P.: High-order neural network structures for identification of dynamical systems. IEEE Trans. on Neural Netw. **6**(2), 422–431 (1995)
34. Amin, M.F., Islam, M.M., Murase, K.: Ensemble of single-layered complex-valued neural networks for classification tasks. Neurocomputing **72**(10–12), 2227–2234 (2009)
35. Savitha, R., Suresh, S., Sundararajan, N., Kim, H.J.: Fast learning fully complex-valued classifiers for real-valued classification problems. In: Liu, D., et al. (ed.). ISNN 2011, Part I, Lecture Notes in Computer Science (LNCS), vol. 6675, pp. 602–609 (2011)

Chapter 2
Neurocomputing with High Dimensional Parameters

Abstract Neurocomputing has established its identity for robustness toward ill-defined and noisy problems in science and engineering. This is due to the fact that artificial neural networks have good ability of learning, generalization, and association. In recent past, different kinds of neural networks are proposed and successfully applied for various applications concerning single dimension parameters. Some of the important variants are radial basis neural network, multilayer perceptron, support vector machines, functional link networks, and higher order neural network. These variants with single dimension parameters have been employed for various machine learning problems in single and high dimensions. A single neuron can take only real value as its input, therefore a network should be configured so that conventionally use as many neurons as the dimensions (parameters) in high dimensional data for accepting each input. This type of configuration is sometimes unnatural and also may not achieve satisfactory performance for high dimensional problems. It has been revealed by extensive research work done in recent past that neural networks with high dimension parameters have several advantages and better learning capability for high dimensional problems over conventional one. Moreover, they have surprising ability to learn and generalize phase information among the different components simultaneously with magnitude, which is not possible with the conventional neural network. There are two approaches to naturally extend the dimensionality of data elements as single entity in high dimensional neural networks. In first line of attack the number field is extended from real number (single dimension) to complex number (two dimension), to quaternion (four dimension), to octanion (eight dimension). The second tactic is to extend the dimensionality of data element using high dimensional vector with scalar components, i.e., three dimension and N-dimension real-valued vectors. Applications of these numbers and vectors to neural networks have been extensively investigated in this chapter.

2.1 Neuro-Computing with Single Dimensional Parameters

In recent years, neurocomputing have emerged as a powerful technique for various tasks such as function approximation, classification, clustering, and prediction in wide spectrum of applications. Multilayer neural network and back-propagation

B.K. Tripathi, *High Dimensional Neurocomputing*,
Studies in Computational Intelligence 571, DOI 10.1007/978-81-322-2074-9_2

learning algorithm for its training are most popular in neural networks community. The primary aim of the neural network is to learn input to output mapping, and the learning algorithm achieves it by adjusting the parameters of the network, which are weights and threshold values. As these weights and thresholds are real values in the conventional neural network, it is also called real valued neural network (RVNN) or conventional ANN or refers to neurocomputing with single dimension parameters.

A conventional ANN is a model that apes the real neuron described by many researchers in the model description suggested time to time. The artificial neurons are shown connected with links going from one layer to the one immediately succeeding it, and some applications of neural networks, however, have had synapses that link the neurons of the present layer with the ones not only of the immediately succeeding layer, but also to the neurons that lie further up in the line (Lang and Witbrock 1988). The strength of the synapses (connections) is the synaptic strength, is the weight associated with the connection. In the human brain, the weight is actually the potential that controls the flow of electric impulses through the link. Each neuron has a well defined aggregation function to process the integration of impinging signals and an activation function that limits the output in predefined range. A typical activation function that closely resembles the activation of real biological neuron is sigmoid function as shown in Fig. 2.1. A steepness factor was introduced to adjust the shape of the activation function and tailor it to a form that closely resembles the actual characteristic. A number of neurobiological studies and biophysics of computation have inspired researchers to precisely state different aggregation function in literature. Chapter 4 of this book presents three new higher order neuron models based on these studies.

ANN in real domain have limitations such as slow convergence and degree of accuracy achieved is normally lower, specially for many applications, which deal with high dimensional signals. The easiest solution would be to consider a conventional real domain neural network, where high dimensional signals are replaced by inde-

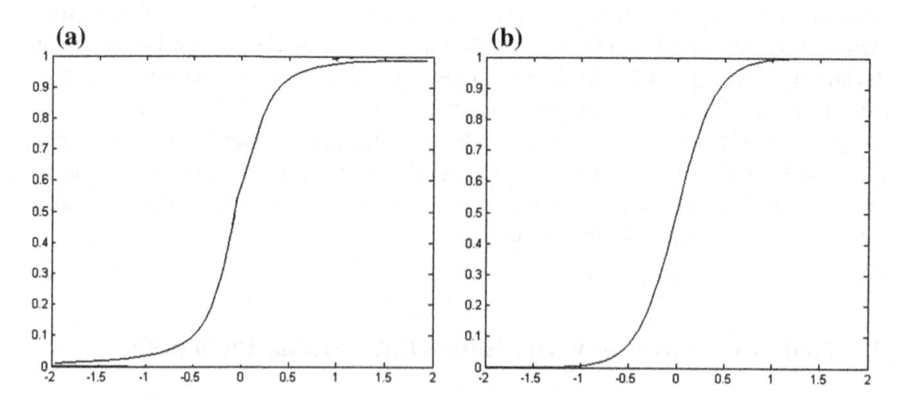

Fig. 2.1 Comparing the actual response *curve* of biological neuron with the mathematical function. **a** The actual characteristic at the output of the neuron in the brain. **b** The characteristic described by sigmoid function

pendent real-valued signals. Such a real-valued neural network is unable to perform mapping in higher dimensions because corresponding learning algorithms cannot preserve each input point's angle in magnitude as well as in sense. Besides that the huge network topology is another limiting factor, which enhances storage memory requirement, If conventional ANN is applied for high dimensional problems, they also require a large number of training iterations for the acceptable solutions and provide poor class distinctiveness in classification. There may be cases in which learning speed is a limiting factor in practical applications of neural networks to problems that require high accuracy. The most acceptable solution resulted from researches is to consider neural network designed with high dimensional parameters. High dimension neural networks have been found worth while in recent researches to overcome from these problems considerably. This book is an attempt to further improve many issues through consideration of theoretical and practical aspects of high dimensional neurocomputing.

2.2 Neurocomputing in High Dimension

In our day to day life, we come across many quantities that involve only one value (magnitude), which is a real number. However, there are also many quantities that involve magnitude and direction. Such quantities are generally called vectors, which may be represented by hypercomplex number system and/or real-valued vectors. It is a matter of universal incidence that a vector represents a cluster of particles in the factual world. The recent researches in neurocomputing are dedicated to formulate a model neuron that can deal with N signals as one cluster, called N-dimensional vector neuron. In this book, high dimensional neurons are defined through these vectors, and high dimensional neural networks such as complex neural networks, quaternary neural networks, and three-dimensional exterior neural networks are unified in terms of a vector representation. These vectors (signals), which are supposed to flow through a high dimensional neural network, are the unit of learning. Therefore, they are capable of learning high-dimensional motions as its inherent property, which is not possible in real domain neural networks.

In science and engineering, we frequently come across with both types of quantities. In mathematics, a hypercomplex number is a traditional term for an element of algebra over a Field.[1] The hypercomplex numbers are the generalization of the concept of real numbers to n dimensions which come up with an unexpected outcome. Our idea of "number like" behavior in $\boldsymbol{R}^n$ is motivated by the cases $n=1$ (real numbers $\boldsymbol{R}$), 2 (complex numbers $\boldsymbol{C}$) that we already know. In trying to generalize the real number to higher dimensions, we find only four dimensions where the

[1] In abstract algebra, a field is a set F, together with two associative binary operations, typically referred to as addition and multiplication.

idea works: n = 1, 2, 4, 8. These number systems have many common algebraic and geometric properties.

A literature survey into the vector algebra brings out the fact that many searches had gone to enable the analysis of quantities, which involve magnitude and direction in three-dimensional space, in the same way as complex numbers had enabled analysis of two-dimensional space, but no one could arrive at a complex numbers like system. Therefore, another way to represent a vector in three dimensional space (or $\boldsymbol{R}^3$) are identified with triples of scalar components. These quantities are often arranged into a real-valued vectors, particularly, when dealing with matrices. These standard basis vectors can also be generalized into n-dimensional space (or $\boldsymbol{R}^n$).

This chapter is devoted to present a theoretical foundation of hypercomplex number system and high dimensional real-valued vector. This chapter also establishes their basic concepts for vector representation along with their algebraic and geometric properties in view of designing high dimensional neural networks. The successive chapters of the book lead to their vital applicability in various areas.

2.2.1 Hypercomplex Number System

The nineteenth century is observed as the very thrilling time for philosophy of complex numbers. Though, complex numbers had been discussed in works published in the sixteenth century, the study of complex numbers was totally dismissed as worthless at that time. After three centuries the sensibleness of complex numbers was truly understood when most of the fundamental results, which now form the core of complex analysis, were exposed by Cauchy, Riemann along with many others. Even Irish mathematician and physicist William Rowen Hamilton was fascinated by the role of complex number system in two-dimensional geometry in the nineteenth century before discovering quaternions. Indeed, set of real numbers '$\boldsymbol{R}$' is a subset of set of complex numbers '$\boldsymbol{C}$'; $\boldsymbol{C}$ is a Field extension over $\boldsymbol{R}$. It is a number system where we can add, subtract, multiply, and divide. The algebraic structure "doublets" or "couplets" ($a + ib \in R^2$) was regarded as a algebraic representation of the Complex Numbers, which can easily use complex arithmetic to do various geometric operations. The Field of complex numbers is defined by

$$\boldsymbol{C} = \{a + i\ b \mid a, b \in R;\ i^2 = -1\}$$

The field of complex numbers is a degree two field extension over the field of real numbers; the vector space of C forms the basis $\{\mathbf{1}, \mathbf{i}\}$ over $\boldsymbol{R}$. This means that every complex number can be written in the form $a + \mathbf{i}\ b$, where a and b are real numbers and i is an imaginary unit ($i^2 = -1$). Mathematicians want to construct a new fields for hypercomplex numbers such that $\boldsymbol{C}$ becomes a subset of hypercomplex number system, and the new operations in them are compatible with the old operations in $\boldsymbol{C}$. Therefore, mathematicians were looking for a field extension of $\boldsymbol{C}$ to higher dimensions, hence hypercomplex numbers.

As next natural step, Hamilton was eager to extend the complex numbers to a new algebraic structure with each element consisting of one real part and two distinct imaginary parts, which would be known as "Triplets" and forms the basis $\{\mathbf{1}, \mathbf{i}, \mathbf{j}\}$. There are necessary and sufficient mathematical reasons, why one should attempt for such a construction. He desired to use these triplets to operate in three-dimensional space, as complex numbers were used to define operations in the two-dimensional plane. He attempted for years to invent an algebra of triplets ($a + \mathbf{i}\, b + \mathbf{j}\mathbf{c} \; \in R^3$) to play same role in three dimensions.

Can triplets be multiplied? Hamilton worked unsuccessfully in creating this structure for over 10 years. After a long work he observed that they can only be added and subtracted; Hamilton could not solve the problem of multiplication and division of triplet. We now know that this pursuit was in vain. Historically, it was noted that on October 16th, 1843, while walking with his wife along the Royal Canal in Dublin, the concepts of using quadruple with the rules of multiplications dawned on him. Hamilton discovered a four dimensional algebraic structure called the quaternions. The discovery of the quaternions is one of the most well documented discoveries in mathematics. In general, it is very rare that the date and location of a major mathematical discovery are known. Hamilton explicitly stated, "I then and there felt the galvanic circuit of thought and the sparks which fell from it were the fundamental equations between $\mathbf{i}, \mathbf{j}, \mathbf{k}$; exactly such as I have used them ever since in complex number system".

The set of quaternions, often denoted by $\boldsymbol{H}$ in honor of its discoverer, constitute a noncommutative field (a skew field) that extends the field $\boldsymbol{C}$ of complex numbers. The quaternions is constructed by adding two new elements $\mathbf{j}$ and $\mathbf{k}$ in complex number, thus new algebraic structure would require three imaginary parts along with one real part. For this new structure to work, Hamilton realized that these new imaginary elements would have to satisfy the following conditions:

$$i^2 = j^2 = k^2 = ijk = -1$$

One could now talk about the additive and multiplicative operations that can be defined on elements of H and turn it into a field. The field of quaternions can then be written as

$$\boldsymbol{H} = \{q = q_0 + q_1\mathbf{i} + q_2\mathbf{j} + q_3\mathbf{k} \mid q_n \in \boldsymbol{R}; \;\; \text{and} \;\; i^2 = j^2 = k^2 = ijk = -1\}$$

If Hamilton had been able to develop his Theory of Triplets, he would have effectively built a degree three field extension of $\boldsymbol{R}$ whose vector space forms the basis $\{\mathbf{1};\ \mathbf{i};\ \mathbf{j}\}$ over $\boldsymbol{R}$ such that $i^2 = j^2 = -1$. This field must be closed under multiplication. After struggling with all possibilities, researchers including himself came in conclusion that there is no third degree Field extension over $\boldsymbol{R}$ with basis $\{\mathbf{1};\ \mathbf{i};\ \mathbf{j}\}$ holding the properties as in $\boldsymbol{C}$. Thus, it is not possible to create the Theory of Triplets while satisfying the requirements of a Field. Hamilton had to abandon the Theory of Triplets.[2] The extensive investigations show why Hamilton had to consider a four dimensional Field extension by adding a new element **k** that is linearly independent of the generators **1**, **i**, and **j**; and whose vector space forms the basis {**1**; **i**; **j**; **k**} such that $i^2 = j^2 = k^2 = -1$.

A quaternion is a hypercomplex number, which is an extension to the complex numbers. The hypercomplex number domain possesses a "Numberlike" behavior in $\boldsymbol{R}^n$ and consists of symbolic expression of n terms with real coefficients, where n may be 1 (real numbers), 2 (complex numbers), 4 (quaternion), 8 (cayley numbers or octonion). These numbers can also be considered as Field extension of classical algebra in hypercomplex number domain. They share many properties with complex numbers with interesting exceptions. Quaternion algebra has all the required properties except commutative multiplication, whereas the octonion algebra has all the required properties except commutative and associative multiplication. They are now used in computer graphics, computer vision, robotics, control theory, signal processing, attitude control, physics, bioinformatics, molecular dynamics, computer simulations, and orbital mechanics.

2.2.2 Neurocomputing with Hypercomplex Numbers

The brief survey into the development of family of hypercomplex numbers point out the fact that the idea of developing these numbers may generate normed division algebras only in dimensions 1 (real numbers), 2 (complex numbers), 4 (quaternions), and 8 (octonions). Hypercomplex numbers are direct extension of the complex number into the high dimension space. They can be seen as high dimensional vectors comprising of components with one scalar and a vector in space. High-dimensional neural networks developed using these numbers has natural ability of learning motion in corresponding dimension because the unit of learning are these numbers (signals) flowing through respective neural network. The neural network in hypercomplex domain is an extension of the classical neural network in real domain, whose weights, threshold values, input, and output signals are all hypercomplex numbers. This chapter will clarify the fundamental properties of a neuron and neurocomputing with hypercomplex number system.

2.2.3 Neurocomputing with Vectors

The word vector was originated from the Latin word vectus, which stands for "to carry". The modern vector theory was evolved from early nineteenth century when an

[2] Interested readers may consult modern abstract algebra to understand difficulty of building a three-dimensional field extension over $\boldsymbol{R}$ and Hamiltons breakthrough concerning the necessity of three distinct imaginary parts along with one real.

interpretation of vector in different dimensions had been given through the tuples of scalar components. Vectors in a three dimensional space (or $\boldsymbol{R}^3$) can be represented as coordinate vectors in a Cartesian coordinate system and can be identified with an ordered list of three real numbers (tuples [x, y, z]). These numbers are typically called the scalar projections (or scalar components) of the vector on the axes of the coordinate system. In wide spectrum, a vector in n-dimensional space (or $\boldsymbol{R}^n$ or spatial vector) is a geometric quantity having magnitude (or length) and direction expressed numerically as n-tuples, splitting the entire quantity into its orthogonal-axis components. The vector-valued neurons were introduced as natural extension of conventional real-valued neuron, which influence the behavioral characteristics of the neuron in high dimensional space. We live in a three dimensional world, certainly all of our movements are in 3D. There are many natural aspects of learning 3-D motion in space particularly through neurocomputing. The purpose of 3D vector-valued neural network to 3D geometry is that it makes the neurocomputing study simple and elegant.

2.3 Neurocomputing with Two Dimensional Parameters

The neurocomputing has found application in almost all industry and every branch of science and technology. Our understanding of the ANN improved over the years with the much light research in the direction threw, with the lasting contributions of McCulloch and Pitts (1943), Donald Hebb (1949), Minsky (1954), Rosenblatt (1958), Minsky and Papert (1969), Werbos (1974), Fukushima and Miyaka (1980), John Hopfield (1982), Nitta (1997), Adeli (2002), Aizenberg (2007), Tripathi (2010) to name a few. Among the most recent developments in the area are the complex variable based neural networks (CVNN) that represent a second generation of architectures and also scored over the standard real variable based networks (ANN) in certain aspects. All the parameters including synaptic weights, bias, input-output, and signals flowing through network are complex numbers, aggregation, and activation functions are also in complex domain. Since it operates in the complex variables setting, the conventional Back-Propagation Algorithm (BP) that trains the ANN is not suitable to train the CVNN. The operations on functions in complex domain are not as straightforward as in real domain; therefore, variations in extension of the BP to the complex variables was reported by Leung and Haykin (2010) [1], Piazza (1992) [2], Nitta (2000) [3], Aizenberg (2007) [4], Adeli (2002) [5] called the Back-Propagation Algorithm in Complex Domain (*C*BP). It is imperative that we study the ANNs with two-dimensional parameters with a view to investigate how the new tools of approximation perform in comparison with the existing ones.

In order to preserve the relationship between phase and magnitude in signals, one certainly requires a mathematical representation; and this representation is only possible in the domain of complex numbers. Hence, the model representation of the systems involving these signals should deal with complex values rather than real values. This indicates that the complex variable-based neural network may be useful for such applications. CVNN is the extension of RVNN, in which all the parameters and signals flowing through it are complex numbers (Two Dimension Parameters) in contrast to real numbers in the RVNN. The different neurobiological studies revealed that the action potential in human brain may have different pulse patterns and the distance between pulses may be different. This justifies the introduction of complex numbers representing phase and magnitude by a single entity into neural networks. CVNN has been applied to various fields like adaptive signal processing, speech processing, and communicating, but are not limited. These applications often use signals, which have two types of information embedded in it, the magnitude and the phase. In real domain it is not possible to represent both these quantities by a single quantity; so the magnitude and phase is represented by two numbers and then the neural network can be trained using these two quantities as separate inputs. While the neural network trained on this topology may give satisfactory results, but the relationship between phase and magnitude of a signal cannot be represented by the model because of their separation. Chapter 5 demonstrate that the CVNN can learn and generalize linear and bilinear transformation over typical geometric structures on plane. These transformations cannot be learned using RVNN. The CVNN shows excellent generalization capabilities for these transformations because representation of magnitude and phase by single quantity i.e. complex number. It also worth to mention here that CVNN has yielded far better result even in case of real-valued problem (Chap. 4), hence outperformed over equivalent real-valued neural network.

2.3.1 Properties of Complex Plane

The complex plane is the geometric analog of complex numbers, which has a long and mathematically rich history. It is the set of dual numbers over the real that possesses one to one correspondence with the points of cartesian plane. The complex plane is unlike real line, for it is two-dimensional with respect to real numbers and one-dimensional with respect to the set of complex numbers (Halmos 1974). A point on the plane can be viewed as a complex number with the x and y coordinates regarded as the real and imaginary parts of the number. The set of complex numbers is a Field equipped with both addition and multiplication operations, and hence makes a perfect platform of operation. But the order that existed on the set of real numbers is absent in the set of complex numbers, and as a result, no two complex numbers could be compared as being big or small with respect to each other, but their magnitudes (which are real numbers) could well be compared. The properties of the complex plane are different from those of the real line. The set of real numbers was one-dimensional, while as was pointed out, the set of complex numbers is one-dimensional if the field in question is the set of complex numbers itself, while it is two-dimensional if the filed is the set of real numbers. The complex numbers have a magnitude associated with them and a phase that locates the complex number uniquely on the plane. It is hence clear that the learning algorithm on complex plane ($\boldsymbol{C}$BP) that trains the CVNN must not only obtain a convergence with respect to the magnitude, but also with respect to the phase. This is equivalent to stating that the real as well as imaginary parts of the complex numbers must be separately captured by the $\boldsymbol{C}$BP.

2.3.1.1 Beauty of Complex Numbers

Complex domain ($\boldsymbol{C}$) itself is gaining more attention because many real applications involve signals that are inherently complex-valued. A complex number is directly related to the two-dimension data. It comprises of two real numbers and comes with phase information embedded into it. Addition and multiplication are much easier in $\boldsymbol{C}$. Any complex number has a length and angle, hence forms a plane. Their operations are very related to two-dimensional geometry where one can use complex arithmetic to do various geometric operations. Thus, it is more significant in problems where we wish to learn and analyze signal amplitude and phase precisely. Let $\boldsymbol{C}$ be the set of complex numbers and triplet (F, $\bullet$, $\otimes$) be a Field equipped with operations $\bullet$, $\otimes$, satisfying the Closure, Commutative, Associative, Identity, Distribute ($\otimes$ distributes over $\bullet$), and Inverse properties for arbitrary elements belonging to F. A two-dimensional Field with basis $\{\mathbf{1}, \mathbf{i}\}$ forms a two-dimensional vector space of $\boldsymbol{C}$ over $\boldsymbol{R}$.

Definition 2.1 The triplet (F, $\bullet$, $\otimes$) is said to be a Field equipped with operations $\bullet$, $\otimes$; $\forall c_1, c_2, c_3 \in F$, if it satisfies the following axioms with respect to $\bullet$:

1. *Closure* If $c_1, c_2 \in F \Longrightarrow c_1 \bullet c_2 \in F$.
2. *Commutative* $c_1 \bullet c_2 = c_2 \bullet c_1$.
3. *Associative* $c_1 \bullet (c_2 \bullet c_3) = (c_1 \bullet c_2) \bullet c_3$.
4. *Identity* $\exists$ an element called '0' such that $c_1 \bullet 0 = 0 \bullet c_1 = c_1$.
5. *Inverse* $\exists$ a unique c_1 for every c_2 such that $c_1 \bullet c_2 = 0$.

and following axioms with respect to $\otimes$:

1. *Closure* If $c_1, c_2 \in F \Longrightarrow c_1 \otimes c_2 \in F$.
2. *Commutative* $c_1 \otimes c_2 = c_2 \otimes c_1$.
3. *Associative* $c_1 \otimes (c_2 \otimes c_3) = (c_1 \otimes c_2) \otimes c_3$.
4. *Identity* $\exists$ an element called '1' such that $c_1 \otimes 1 = 1 \otimes c_1 = c_1$.
5. *Inverse* $\exists$ a unique c_1 for every $c_2 (\neq 0)$ such that $c_1 \otimes c_2 = 1$.

and also *Distributive Property*, $\otimes$ distributes over $\bullet$ such that $c_1 \otimes (c_2 \bullet c_3) = c_1 \otimes c_2 \bullet c_1 \otimes c_3$.

Definition 2.2 The Field of complex numbers is a degree two field extension over the field of real numbers. The set of complex numbers is a Field equipped with operations $+$, $\times$ such that for every two elements $a + jb$, $c + jd$:

- $(a + jb) + (c + jd) = (a + c) + j(b + d)$.
- $(a + jb) \times (c + jd) = (ac - bd) + j(bc + ad)$.

where j is the imaginary unit defined by the equation $j^2 = -1$. It can be verified that the definitions for $+$, $\times$ satisfy all the postulates of the Field.

2.3.1.2 Cauchy-Riemann Equations and Liouville's Theorem

In real domain, the property of differentiability is not a very strong property for functions of real variables. It is surprisingly true that study of complex function for differentiability (analyticity) is a different topic from real analysis. The power and importance of complex numbers cannot be exploited until a full theory of analytic (holomorphic) function is developed. Interested readers may consult the theory of complex numbers for details, this section only presents the brief discussion on differentiability necessary for the development of learning algorithm in neurocomputing.

Definition 2.3 A complex valued function $f\colon \boldsymbol{C} \longrightarrow \boldsymbol{C}$ is said to be ***analytic*** (complex differentiable) at $z \in \boldsymbol{C}$ if the following limit exists at every point z in the complex plane. If the function is analytic over the whole finite complex plane, it is said to be an entire function.

$$\text{Limit}_{\delta z \to 0} \frac{f(z+\delta z) - f(z)}{\delta z} \tag{2.1}$$

The definition demands that the function be differentiable at every point in some neighborhood of the point z. The function f is said to be differentiable at z when its derivative at z exists. The limit will be called the first derivative of f at z and denoted by $f'(z)$.

Further, there are a pair of equations that the first-order partial derivatives of the component functions of a function (f) with complex variable, must satisfy at a point when the derivative of f exists there.

Let $f(z) = U(x, y) + jV(x, y)$ be a complex valued function. Cauchy-Riemann equations (***CR***) are pair of equations that first-order partial derivatives of the component functions, U and V, of function f must satisfy at a point when the derivative of f exist there.

Definition 2.4 The first order partial derivatives of component function of f must exist. A complex valued function $f(z) = U(x, y) + jV(x, y)$ is said to satisfy ***CR*** equations if the following equalities hold:

$$\frac{\delta U}{\delta x} = \frac{\delta V}{\delta y} \quad \text{and} \quad \frac{\delta U}{\delta y} = -\frac{\delta V}{\delta x} \tag{2.2}$$

It can be shown that every analytic function satisfies CR equations. The converse is true if an additional condition of continuity of the partial derivatives of the CR is assumed (Ahlfors 1979).

Above equations, not only give the derivative of f in terms of partial derivatives of component function, but also they (***CR*** equations) are necessary conditions for the existence of the derivative of a function at a point z. They can be used to locate points at which function does not have a derivative. These equations are not sufficient to ensure the existence of the derivative of a function at that point.

The complex plane unlike the real line is a two-dimensional space. The second dimension adds flexibility and at the same time restricts the choice of activation functions for neural network applications by imposing certain constraints. More precisely, the important constraint imposed by the complex plane is epitomized in the Liouville Theorem.

Theorem 2.1 *The theory of functions in complex domain imposes its own constraint in the form of Liouville Theorem, which states that if a function in complex domain is both entire and bounded, it must be a constant function. As a ramification of the theorem, the constraints emerge:*

- *No analytic function except a constant is bounded in the complex plane.*
- *An analytic complex function cannot be bounded on all points of the complex plane unless it is constant.*

Analyticity, boundedness are the parameters of concern in the search for complex activation functions. The term regular and holomorphic are also interchangeably used in the litrature to denote analyticity. In view of theorem, if a nontrivial complex-valued function is analytic it must go unbounded at least one point on the complex plane, and if the function is bounded it must be nonanalytic in some region for it to qualify as activation function. Hence, a search for activation function should make sure these conditions are satisfied. The second dimension of the complex plane necessitates a study of three-dimensional surfaces (Chap. 3), as the real and imaginary parts of the complex activation functions are both functions of real and imaginary parts of the variable.

Definition 2.5 If a function fails to be analytic at a point z_0, but is analytic at some point in every neighborhood of z_0, then z_0 is called a singular point or singularity of function.

2.3.2 Complex Variable Based Neural Networks

Complex numbers form a superset of real numbers, an algebraic structure that defines real world phenomenon like signal magnitude and phase. These are useful in analyzing various mathematical and geometrical relationships in two dimension space. For nearly a decade, the extension of real-valued neurons for operation on complex signals [2, 5–7] has received much attention in the field of neural networks. The main motivation in designing complex variable based neural networks is to utilize the promising capabilities of complex numbers. Complex numbers are a subfield of quaternions. The decision boundary of the complex-valued neuron consists of two hypersurfaces, which intersect orthogonally each other and divides a decision region into 2^2 (=4) equal sections. Now, neural network dealing with complex numbers are not the new entrants to the field of neural networks; they have established the basic theories, yet they require more exploration in new coming applications. A brief survey into the CVNN brings out the fact that it provides faster convergence with better results, reduction in learning parameters (network topology), and the ability to learn two-dimensional motion of signal [3, 8]. The weight update rule for the CVNN is exactly same as the one used to training networks using the most popular error correction learning.

$$w(t+1) = w(t) + \eta \frac{\delta E)}{\delta w(t)} \tag{2.3}$$

where w is the weight that get updated as the algorithm runs iteratively, E is the Error Function that gets minimized in the process of weight update and η is the learning rate. The difference of course lies in the fact that the weights are complex numbers, while the error function and learning constant are positive real numbers. All the signals that the neurons fire in response to aggregation and activation functions are all complex in nature. It must be emphasized here that continuity and differentiability of a function in complex domain play a central role in development of complex variable based neural networks. It is hence obvious that a thorough discussion of complex variables and complex mappings is essential to comprehend the mechanism by which a CVNN operates. The definitions discussed here are required to prepare the ground for a systematic study to develop a theory for analysis.

2.4 Neurocomputing with Three Dimensional Parameters

Artificial neural networks have been studied for many years in the hope of achieving human like flexibility in processing typical information. Some of the recent researches in neurocomputing concerns the development of neurons dealing with three-dimensional parameters [9] and their applications to the problems, which deals with three-dimensional information. There has been rapid development in the field of 3D imaging, computer vision, and robotics in last few years. These are multidisciplinary fields, which encompasses various research areas and deal with information processing through modern neurocomputing paradigm. They are at their infancy [10–12] and requires to explore methods based on neural networks. The 3D motion interpretation and 3D feature recognition are essential part of high level analysis, and found wide practical uses in the system development of these fields. Although, there are many methodologies [10, 11, 13, 14] to solve them, they instead use extensive mathematics and are time consuming. They are also weak to noise. Therefore, it is desirable for realistic system to consider iterative methods, which can adapt system for three-dimensional applications. This book is aimed at presenting relevant theoretical and experimental framework based on multilayer neural networks of 3D vector-valued neurons. In Chap. 6, we present a straightforward technique that uses 3D geometric point set (point cloud) representation of objects. The method described here is fully automatic, does not require much preprocessing steps, and converges rapidly to a global minima.

2.4.1 Properties of Vector Space

A vector space is a set whose elements are called vectors. The vector space is intuitively spatial since all available directions of motion can be plotted directly onto a spatial map. The idea of a vector is far more general than the picture of a line with an arrowhead attached to its end. In general, a vector is thought a directed arrow pointing from the origin to the end point given by the list of numbers. A vector is a list of numbers, and the dimensionality of a vector is length of the list, where each number represents the vector's component in the dimensions. A 3-dimensional vector would be a list of three numbers, and they live in a 3-D volume. Vector space offers a convenient way to describe different geometric properties. There are two operations (addition and scalar multiplication) defined on them, which must obey certain simple rules, the axioms for a vector space.

Definition 2.6 Let set $\boldsymbol{V}$ be the vector space and the tuple $(F, +)$ is a Field equipped with operations $+$. The $\boldsymbol{V}$ equipped with the operation '+' is said to be a vector space over the field F if the following axioms are satisfied $(\forall v_1, v_2, v_3 \in V)$

- *Closure* vector space is closed under addition and multiplication by scalars $v_1 + v_2 \in \boldsymbol{V}$ and $\alpha\, v_1 \in \boldsymbol{V}$ where α is a scalar.
- *Commutative* The commutative law of addition holds $v_1 + v_2 = v_2 + v_1$.
- *Associative* $v_1 + (v_2 + v_3) = (v_1 + v_2) + v_3$.
- *Identity* There is a zero vector, so that for each vector, $\exists$ an element called '0' such that $v_1 + 0 = v_1$.
- *Inverse* There is a unique additive inverse for each vector, for each vector v_1, $\exists$ another vector v_2 such that $v_1 + v_2 = 0$; then $v_2 = -v_1$.

2.4.2 3D Vector Based Neural Networks

In machine learning the interpretation of 3D motion, 3D transformations, and 3D object matching are few expected applications. Though, there have been many methodologies to solve them, however, these methods are time consuming and weak to noise. This book presents an efficient solution using multilayered network of 3D vector-valued neurons. In 3D vector valued neural network the parameters like threshold values, input-output signals are all 3D real valued vectors, and weights associated with connections are 3D orthogonal matrices. All the operations in such neural models are scaler matrix operations. The corresponding 3D vector valued back-propagation algorithm (3DV-BP) is a natural extension of complex valued back-propagation learning algorithm [2, 8], and has the ability to learn *3D motion* as complex-BP learn 2D motion as its inherent property. In this book, the author investigates the characteristics of 3D vector-valued neural networks by various computational experiments. The experiments suggest that 3DV-BP networks can approximate 3D mapping just by training them only over a part of the domain of the mapping. Chapter 6 explains the learning rule for 3D vector-valued neural networks. The generalization ability of 3D neural network in 3D motion interpretation and in 3D face recognition applications is confirmed through diverse test patterns in Chap. 6.

2.5 Neurocomputing with Four-Dimensional Parameters

The four-dimensional hypercomplex numbers, the Quaternions, have been extensively employed in several fields, such as modern mathematics, physics, control of satellites, computer graphics, etc. One of the benefits in graphics provided by quaternions is that affine transformations (especially spatial rotations) of geometric constructs in three-dimensional spaces, can be represented compactly and efficiently. How we should treat data with four-dimension in artificial neural networks? Although this problem can of course be solved by applying several real-valued or complex-valued neurons. But, a better choice may be to introduce a four-dimensional hypercomplex number system based neural network, that could be confronted to the Quaternion in the same way as the complex numbers are confronted to the CVNN. This hypercomplex number system was introduced by Hamilton [15], which treat four-dimensional data elements as a single entity. There has been a growing number of interests concerning the use of neural networks in the quaternionic domain [16]. All variables in the multilayered quaternionic-valued neural network (QVNN), such as input, output, action potential, and connection weights are encoded by quaternions. A quaternionic equivalent of error back-propagation algorithm has also been investigated and theoretically explored by many researchers. The derivation of this learning scheme adopted a famous Wirtinger calculus [17] because this calculus enables a more straightforward derivation of learning rules.

2.5.1 Properties of Quaternionic Space

The quaternionic space is a four-dimensional vector space over the real numbers. Since the quaternionic algebra is at infancy there are many representations of it, which leads to variations in operations and properties. Hence, demands for wide consensus among researchers in quaternionic space. Most of the researchers have followed Wirtinger calculus in their basic constructions. Representing quaternion as a vector is more compact as well as intuitively straightforward. The quaternions may also be used for three-dimensional operations assuming 3D space as being pure imaginary quaternion.

2.5.1.1 Beauty of Quaternionic Numbers

A quaternion, the generalization of complex number, is a hypercomplex number where complex analysis would be self evident within the structure of quaternion analysis. Unlike the complex number, the quaternion has four components: one is real and the other three are all imaginary.

Definition 2.7 A class of hypercomplex numbers, the quaternions, are defined as a vector **q** in a four-dimensional vector space over the real numbers ($\boldsymbol{R}$) with an ordered basis. Each number is a quadruple consisting a real number and three imaginary numbers i, j, and k. A quaternion $q \in \boldsymbol{H}$ is expressed by fundamental formula

$$\mathbf{q} = q_r + q_i\,\mathbf{i} + q_j\,\mathbf{j} + q_k\,\mathbf{k} \tag{2.4}$$

where q_r, q_i, q_j and q_k are real numbers. The set of quaternions $\boldsymbol{H}$, which is equal to R^4, constitutes a four dimensional vector space over the real numbers with basis $\{\mathbf{1}; \mathbf{i}; \mathbf{j}; \mathbf{k}\}$.

Definition 2.8 The quaternion $\mathbf{q} \in \boldsymbol{H}$ can also be interpreted as having a real part q_r and vector part $\bar{q}$, where the elements $\{\mathbf{i}, \mathbf{j}, \mathbf{k}\}$ are given an added geometric interpretation as unit vectors along the X, Y, Z axis respectively. Equation 2.4 can also be written using 4-tuple or 2-tuple (one scalar and one vector in three space) notation as

$$\mathbf{q} = (q_r,\ q_i,\ q_j,\ q_k) = (q_r, \bar{q}) \tag{2.5}$$

where $\bar{q} = q_i,\ q_j,\ q_k$. Accordingly, the subspace $\mathbf{q} = q_r + 0\,\mathbf{i} + 0\,\mathbf{j} + 0\,\mathbf{k}$ of quaternions may be regarded as being equivalent to the real numbers. The subspace $\mathbf{q} = 0 + q_i\,\mathbf{i} + q_j\,\mathbf{j} + q_k\,\mathbf{k}$ may be regarded as being equivalent to the ordinary 3D vector in R^3.

Definition 2.9 In Wirtinger calculus, a quaternion and its conjugate are treated as independent of each other, which makes the derivation of the learning scheme in neural network clear and compact. The quaternion conjugate is defined as

$$\mathbf{q}^{\aleph} = (q_r,\ -\bar{q}) = q_r - q_i\,\mathbf{i} - q_j\,\mathbf{j} - q_k\,\mathbf{k} \tag{2.6}$$

Definition 2.10 According to the Hamilton rule the quaternion basis satisfy the following identities, which immediately follows that multiplication of quaternions is not commutative.

$$\mathbf{i}^2 = \mathbf{j}^2 = \mathbf{k}^2 = \mathbf{ijk} = -1 \tag{2.7}$$

$$\mathbf{ij} = -\mathbf{ji} = \mathbf{k};\ \mathbf{jk} = -\mathbf{kj} = \mathbf{i};\ \mathbf{ki} = -\mathbf{ik} = \mathbf{j}; \tag{2.8}$$

Definition 2.11 Let $\mathbf{p} = (p_r, \bar{p})$ and $\mathbf{q} = (q_r, \bar{q})$. Customarily, the extension of an algebra attempts to preserve the basic operations defined in the original algebra.

Let $\bar{p} \cdot \bar{q}$ and $\bar{p} \times \bar{q}$ denote the dot and cross products between the three dimensional vectors $\bar{p}$ $\bar{q}$. Then basic operations between quaternions can be defined as follows:

- The addition and subtraction of quaternions are defined in a similar manner as for complex-valued numbers:

$$\mathbf{p} \pm \mathbf{q} = (p_r \pm q_r,\ \bar{p} \pm \bar{q}) = (p_r \pm q_r,\ p_i \pm q_i,\ p_j \pm q_j,\ p_k \pm q_k) \quad (2.9)$$

- The product of p and q is determined using Eq. 2.8 as

$$\mathbf{p}\ \mathbf{q} = (p_r\, q_r - \bar{p} \cdot \bar{q},\ p_r\, \bar{q} + q_r\, \bar{p} + \bar{p} \times \bar{q}) \quad (2.10)$$

- The conjugate of the product is defined as

$$(\mathbf{pq})^{\aleph} = \mathbf{q}^{\aleph}\mathbf{p}^{\aleph} \quad (2.11)$$

- The quaternion norm of $\mathbf{q}$, denoted by $|\mathbf{q}|$, is defined as

$$|\mathbf{q}| = \sqrt{\mathbf{q}\mathbf{q}^{\aleph}} = \sqrt{q_r^2 + q_i^2 + q_j^2 + q_k^2} \quad (2.12)$$

2.5.1.2 Cauchy-Riemann-Fueter Equation

The Swiss mathematician Fueter developed the appropriate generalization of the Cauchy-Riemann equations to the quaternionic functions. The analytic condition for the quaternionic functions is defined by Cauchy-Riemann-Fueter (CRF) equation, which corresponds as an extension of the Cauchy-Riemann (CR) equations defined for the functions in complex domain. In order to construct learning rules for quaternionic neural networks, CRF equation describes the required analyticity (or differentiability) of the function in the quaternionic domain.

Definition 2.12 Let f: $\boldsymbol{H} \rightarrow \boldsymbol{H}$ be a quaternionic valued function defined over a quaternionic variable. The condition for differentiability of any quaternionic function f is defined as follows:

$$\frac{\partial f(\mathbf{q})}{\partial q_r} = -\mathbf{i}\frac{\partial f(\mathbf{q})}{\partial q_i} = -\mathbf{j}\frac{\partial f(\mathbf{q})}{\partial q_j} = -\mathbf{k}\frac{\partial f(\mathbf{q})}{\partial q_k} \quad (2.13)$$

An analytic function can serve as the activation function in the neural network. The analytic condition for the quaternionic function, called the Cauchy-Riemann-Fueter (CRF) equation, yields:

$$\frac{\partial f(\mathbf{q})}{\partial q_r} + \mathbf{i}\frac{\partial f(\mathbf{q})}{\partial q_i} + \mathbf{j}\frac{\partial f(\mathbf{q})}{\partial q_j} + \mathbf{k}\frac{\partial f(\mathbf{q})}{\partial q_k} = 0 \quad (2.14)$$

CRF equation is necessary condition to assure analyticity in the quaternionic domain.

2.5.2 Quaternionic Activation Function

A major issue in designing neural networks in the quaternionic domain is about the introduction of suitable functions for the activation in updating the neurons' states. It is worthwhile to consider the capability of the quaternionic neural network with standard activation functions, as in CVNN, i.e., whether the considered network can approximate with given functions. A popular approach of activation function in complex domain, the so-called "split" type function,[3] has also been applied by researchers in many applications where a real-valued function is applied to update each component of a quaternionic value. Real-valued sigmoidal function and hyperbolic function, which are differentiable, are often used for this purpose. However, due to lack of analyticity in split-type quaternionic function many, others researchers suggested it an inappropriate approach.

The CRF equation has pointed out another issue on the standard analyticity of function in the quaternionic domain. Only linear functions and constants satisfy the CRF equation. For example, an analytic quaternionic tanh function can be used as an activation function, but this function may contain several kinds of singularities as in the case of corresponding function in complex domain, and hence unbounded. Thus, this quaternionic neural network may face the problem of the existence of singularities. Yet this problem can also be handled similarly as in the case of CVNN, through the removal or avoidance of such singularities; which can not considered a better choice.

Recently, another class of analyticity, called "local analyticity", has been developed for the quaternionic functions [18], and is distinguished from the standard or global analyticity. This analytic condition is derived at a quaternionic point with its local coordinate, rather than in a quaternionic space with a global coordinate. Interested reader my consult [18, 19] for details. The derivation of local analytic condition shows that a quaternion in the local coordinate system is isomorphic to the complex number system, and thus it can be treated as a complex value. A neural network with an activation function with local analyticity has been first proposed and analyzed by Mandic (2011), in [20]. Such networks demonstrated their outperformance over the network with a split-type activation function in several applications.

[3] Split complex function refers to functions $f\colon \boldsymbol{C} \longrightarrow \boldsymbol{C}$ for which the real and imaginary part of the complex argument are processed separately by a real function of real argument.

2.5.3 Quaternionic Variable Based Neural Networks

Quaternary neural networks were proposed by Arena and Nitta independently in the mid-1990s [21] The architecture of the multilayered QVNN is same as conventional neural networks except that the neurons, which process quaternionic signals and parameters in the network, are encoded by quaternionic values [16]. A quaternionic activation function introduces nonlinearity between the action potential and output in the neuron. One can treat quaternionic functions in the same manner as complex-valued functions, but under the condition of local analyticity. The quaternionic functions with local analytic conditions are isomorphic to the complex functions, thus several activation functions (such as complex-valued sigmoid functions) can be used extendedly in the quaternionic domain. Many researchers have independently considered the capability of the proposed quaternionic network with different activation functions. The connection weights in QVNN may be updated by error back-propagation learning algorithms, until the desired output signals could not be obtained with respect to the input signals. In 1927, Wirtinger calculus [17] was basically invented for analysis of complex numbers, but now it appears to be prominent in quaternionic algebra, and hence yielded a solid representation for QVNN constructions. Analytic conditions for quaternionic functions may be derived by defining a complex plane at a quaternionic point, which is a kind of reduction from quaternionic domain to complex domain.

A quaternion possesses four degrees of freedom, therefore the decision boundary of the quaternary neuron consists of four hypersurfaces, which intersect orthogonally with each other and divides a decision region into 2^4 (=16) equal sections. Nitta [16] has successfully solved the 4-bit parity problem by a single quaternary neuron with the orthogonal decision boundary, which cannot be solved with a single real-valued neuron, resulting in the highest generalization ability, and hence reveals a potent computational power of the quaternary neuron. The neural networks with quaternionic neurons has been recently explored in an effort to naturally process three or four dimensional vector data, such as color/multi-spectral image processing, predictions for three-dimensional protein structures and controls of motion in high-dimensional space. They can be highly effective in the fields such as robotics and computer vision in which quaternions have been found useful. The application of QVNN to engineering problems, such as color night vision, predictions for the output of chaos circuits, and winds in three-dimensional space is also challenging, but will be the candidates for application.

2.6 Neurocomputing with N-Dimensional Parameters

In many important applications of science and engineering, signals and system parameters are most conveniently represented as a vector in N-dimensional space using an ordered N-tuple $[x_1, x_2, \ldots, x_N]$. High dimensional neural networks designed taking account of task domains has shown their superior computational power in wide spectrum of tasks. Thus, neural network models dealing with N signals as one cluster are desired under the powerful framework of N-dimensional vector neuron. An N-dimensional vector neuron is a natural extension of the 3-dimensional vector neuron whose vital applicability has been presented in Chap. 6. The famous neuroscience researcher T Nitta (2007) proposed an efficient solution for the N -bit parity problem with a single N-dimensional vector-valued neuron [22] considering the orthogonal decision boundary. It reveals the potent computational power of N-dimensional vector neurons because this problem cannot be solved with a single usual real-valued neuron. It is reasonable to emphasize it as a new directionality for enhancing the capability of neural networks, and therefore worth researching the neural networks with N-dimensional vector neuron.

2.6.1 Properties of Vectors in R^N

The vectors in space can be directly extended to vectors in N-space. A vector in N-space is represented by an ordered N-tuples $[x_1, x_2, \ldots, x_N]$ of real numbers and same for a point in N-space, R^N. All the listed axioms in Definition 2.6 including two operations (vector addition and scalar multiplication) holds for any three vectors $v_1, v_2, v_3 \in \boldsymbol{V}$ in N-space (R^N), therefore $\boldsymbol{V}$ is called a vector space over the real numeric's $\boldsymbol{R}$.

2.6.2 N-Dimensional Vector Based Neural Networks

The structure of N-dimensional vector neuron, which can deal with N signals in one cluster, can be given by extending the structure of 3-dimensional vector-valued neuron. In an N-dimensional vector-valued neuron all the input-output signals, thresholds are N-D real-valued vectors and the weights are N-dimensional orthogonal matrices. Additional restrictions imposed on the N-dimensional orthogonal matrix (e.g., it can be regular, symmetric, or orthogonal etc.) will also influence the behavioral characteristics of the neuron. The net potential of a N-dimensional neuron can be given as:

$$Y = \sum_{l=1}^{L} W_l \quad X_l + \theta \tag{2.15}$$

where input signal $X_l = [x_1, x_2, \ldots, x_N]^T$ is lth input signal, W_l is the N-dimensional orthogonal weight matrix for the lth input signal and $\theta = [\theta_1, \theta_2, \ldots, \theta_N]^T$ is the threshold value. It is also important to mention here that the N-dimensional vector neuron presented here may be considered with the traditional activation functions. The output of the neuron will also be a N-dimensional real-valued vector. Similar to complex-valued neuron, the decision boundary of an N-dimensional vector neuron consists of N hyperplanes, which intersect orthogonally with each other and divides a decision region into N equal sections. Minsky and Papert (1969) considered the parity problem the most difficult because output required is 1 if the input pattern contains an odd number of 1 s and 0 otherwise. A solution for N-bit parity problem obtained using a single N-dimensional vector neuron demonstrates its highest generalization ability. It is significant to emphasize here that the rational improvement in the number of learning parameters and the number of layers could be achieved with the N-dimensional vector-valued neuron in solving problems possessing high dimensional parameters.

2.7 Concluding Remarks

The theories in neurocomputing have been developed to build mathematical models that mimic the computing power of the human brain. Their powerful processing capability has been demonstrated in various applications of real domain. Traditional neural networks parameters are real numbers and usually used to deal with single dimension. Still, there are many applications, which deal with high dimensional signals. The easiest solution would be to consider a conventional real domain neural network, where high dimensional signals are replaced by independent real-valued signals. Such a real-valued neural network may be highly complex and unrealistic, and besides such network is unable to perform mapping on a high dimension because corresponding learning algorithms cannot preserve each point's angle in magnitude and sense. An alternative is to introduce a neural network with high-dimensional parameters, which comprises of different components as real numbers, and comes with phase information imbedded into it. This approach yields more efficient solution both in terms of computational complexity and performance. Besides, they overcome the users from huge network topology and large storage requirements, whereas enhances the learning speed.

Another competitive advantages of neuro-computing with high- dimensional parameters is the ease with which they may be applied to poorly understood problems in higher dimensions. Neuron is its basic working unit, which does not have predefined meaning, and it evolves during learning in a manner which can characterize the target function. A high dimensional neural network has natural tendency of acquiring high-dimensional information in training, which include magnitudes and phase in a single entity. They are specially useful in areas, where there is a need of capturing phase information in signals, and must be retained all through the problem. This book is an attempt to investigate the functional capabilities of neurons with high-dimensional parameters. The strength and effectiveness of the high-dimensional neural networks have been extensively justified in successive chapters through simulations on different types of problems viz. classification, function approximation, and conformal mapping.

The processing of high-dimensional data is an important task for artificial neural networks. Multilayered neural networks of different class of high-dimensional parameters are presented and analyzed in this book. All neuronal parameters such as input, output, action potential, and connection weights are encoded by respective high-dimensional number system. The computational capability of a single complex-valued, vector-valued, or quaternionic-valued neuron has been independently presented. In order to construct learning algorithm for respective networks analytic, local analytic, or non-analytic conditions may be imposed on the activation function in updating neuron's states. Instead of using conventional description, i.e., cartesian representation, the Cliff-Ford algebra, Vector Calculus, and Wirtinger calculus may be adopted to standardize the learning rules of high dimensional neural networks.

References

1. Leung, H., Haykin, S.: The complex backpropagation algorithm. IEEE Trans. Sig. Proc. **39**(9), 2101–2104 (1991)
2. Piazza, F., Benvenuto, N.: On the complex backpropagation algorithm. IEEE Trans. Sig. Proc. **40**(4), 967–969 (1992)
3. Nitta, T.: An analysis of the fundamental structure of complex-valued neurons. Neural Process. Lett. **12**, 239–246 (2000)
4. Aizenberg, I., Moraga, C.: Multilayer feedforward neural network based on multi-valued neurons (MLMVN) and a back-propagation learning algorithm. Soft Comput. **11**(2), 169–183 (2007)
5. Kim, T., Adali, T.: Approximation by fully complex multilayer perceptrons. Neural Comput. **15**, 1641–1666 (2003)
6. Hirose, A.: Complex-Valued Neural Networks. Springer, New York (2006)
7. Shin, Y., Keun-Sik, J., Byung-Moon, Y.: A complex pi-sigma network and its application to equalization of nonlinear satellite channels. In: IEEE International Conference on Neural Networks (1997)
8. Nitta, T.: An extension of the back-propagation algorithm to complex numbers. Neural Netw. **10**(8), 1391–1415 (1997)
9. Tripathi, B.K., Kalra, P.K.: On the learning machine for three dimensional mapping. Neural Comput. Appl. **20**(01), 105–111. Springer (2011)
10. Moreno, A.B., Sanchez, A., Velez, J.F., Daz, F.J.: Face recognition using 3D surface-extracted descriptors. In: Proceedings of IMVIPC (2003)
11. Xu, C., Wang, Y., Tan, T., Quan, L.: Automatic 3D face recognition combining global geometric features with local shape variation information. In: Proceedings of AFGR, pp. 308–313 (2004)
12. Chen, L., Zhang, L., Zhang, H., Abdel-Mottaleb, M.: 3D shape constraint for facial feature localization using probabilistic-like output. In: Proceedings of 6th IEEE International Conference on Automatic Face and Gesture Recognition (2004)
13. Achermann, B., Bunke, H.: Classifying range images of human faces with Hausdorff distance. In: Proceedings of ICPR, pp. 809–813 (2000)
14. Blanz, V., Vetter, T.: Face recognition based on fitting a 3D morphable model. IEEE Trans. PAMI **25**(9), 1063–1074 (2003)
15. Hamilton, W.R.: Lectures on Quaternions. Hodges and Smith, Dublin (1853)
16. Nitta, T.: A solution to the 4-bit parity problem with a single quaternary neuron. Neural Inf. Process. Lett. Rev. **5**, 33–39 (2004)
17. Wirtinger, W.: "Zur formalen theorie der funktionen von mehr komplexen ver", anderlichen. Math. Ann. **97**, 357–375 (1927)

18. Leo, S.D., Rotelli, P.P.: Quaternonic analyticity. Appl. Math. Lett. **16**, 1077–1081 (2003)
19. Schwartz, C.: Calculus with a quaternionic variable. J. Math. Phys. **50**, 013523:1013523:11 (2009)
20. Mandic, D., Goh, V.S.L.: Complex Valued Nonlinear Adaptive Filters: Noncircularity, Widely Linear and Neural Models. Wiley, Hoboken (2009)
21. Nitta, T.: A quaternary version of the backpropagation algorithm. In: Proceedings of IEEE International Conference on Neural Networks, vol. 5, pp. 2753–2756 (1995)
22. Nitta, T.: N-dimensional vector neuron. IJCAI Workshop, Hyderabad, India (2007)

Chapter 3
Neurocomputing in Complex Domain

Abstract There are many areas of applications which involve signals that are inherently complex-valued. The characteristics of these applications can be effectively realized if they are operated with the complex-valued neural networks (CVNNs). Apart from that it is also widely observed in researches that the real-valued problems can be solved far efficiently if they are represented and operated in the complex domain. Therefore, CVNNs have emerged a very good alternative in second generation of neurocomputing. The CVNNs to preserve and process the data (signals) in the complex domain itself are gaining more attention over their real-valued counterparts. The use of neural networks is naturally accompanied by the trade-off between issues such as the overfitting, generalization capability, local minima problems, and stability of the weight-update system. The main obstacle in the development of a complex-valued neural network (CVNN) and its learning algorithm is the selection of an appropriate activation function and error function (EF). It can be said that the suitable error function-based training scheme with a proper choice of activation function can substantially decrease the epochs and improve the generalization ability for the problem in question. This chapter presents prominent functions as a basis for making these choices and designing a learning scheme. The choice of EF and activation function in the training scheme also circumvents some of the existing lacunae such as error getting stuck and not progressing below a certain value. This chapter further introduces a novel approach to improve resilient propagation in complex domain for fast learning.

3.1 Complex Domain Neuron

We may recall that complex domain neurons are attractive due to the "reliable theoretical results for their universal approximation abilities and for their generalization power measured by series of researchers" [1]. The complex variable-based neuron differs from the conventional real-valued neuron in almost all respects except of course the architecture. Neurons get activated when signal is impinged, the signals along with the weights and biases are complex numbers. The input to the bias neuron is set to $1 + j$ which operates to enable an offset on complex plane, as is not with the real domain neuron. The Neuron fires according to the function of activation, which in again a complex valued. The properties of the complex plane are much

B.K. Tripathi, *High Dimensional Neurocomputing*,
Studies in Computational Intelligence 571, DOI 10.1007/978-81-322-2074-9_3

different from those of the real line and as a ramification, the convergence obtained with the complex domain learning algorithm differs from the one given by the real domain learning algorithm. Recently, it has been shown that the orthogonal decision boundaries of complex-valued neuron help them to solve classification problems more efficiently than their real-valued counterparts [2]. Since then, several complex-valued classifiers have been developed to solve real-valued classification problems [3–5, 8, 24].

The artificial neural networks in complex domain are gaining more consideration because they offer adjustable strong nonlinear input–output mapping and the superior approximation/classification ability over conventional neural network. Currently, much attention has been paid to convergence and stability evaluation of complex-valued neural network (CVNN) [7]. It also demand awareness for varied energy function and varied activation function-based training rules from users who are not true experts in the neural network field. It must also be noted however that a given data can be tailored in different ways to suit for input to the complex domain neuron. For instance, if in a certain problem there exist some number of inputs and outputs in real domain, a variety of combinations naturally come up in complex domain formulation of problem. This grouping may certainly reduce the data patterns to design a CVNN, thus this is an additional feature of the CVNN that a proper grouping can hasten up the training process and enhance the performance result. As complex numbers are dimension two with respect to the set of real numbers (Halmos 1974), a variety of combinations become available for coupling and so the variety of CVNN designs for a particular problem can be observed [4, 8, 9]. It is hence clear that a number of designs are possible for the data at hand. Needless to say that the interpretation of the simulated result should be done in the way in which the training data were modeled. That is if the output was chosen to be purely complex while training, the imaginary part of the simulated data should be considered for interpretation.

ANN in complex domain can be applied in various fields of science and engineering. In problems like pattern classification, optimization, vector operations, filtering, approximation, 2D motion interpretation and control system applications the principles of CVNN are directly applied. Many problems of actual world are formulated as one of these problems, identifying the relation between the parameters from the physical data with the input–output data and other parameters describing the neural networks. By adopting the introduced approach together with already available theory in the literature, this chapter presents the methodology that optimize the complexity of neural networks; yield efficient learning and better generalization. The chapter is mainly intended to present a comprehensible approach toward stability of the gradient descent weight-update system for complex-valued neural network (CVNNs).

3.1.1 Why Complex Domain Neuron

The conventional neuron (real domain neuron) is difficult to be used for many tasks where data are often represented with values in complex domain. The one easiest solution would be to consider a conventional neuron for learning, where the complex input–output signals are replaced by a pair of independent signals in real domain. But, there is an issue, how to deal with phase information which is well embedded in any complex number or signal? Therefore, the transformation $\boldsymbol{C} \longrightarrow \boldsymbol{R}$, used in the derivation of the learning algorithms, will affect the phase approximation capabilities of these algorithms. For better phase approximation, one needs to use an algorithm which simultaneously minimizes both the magnitude and phase errors [10]. Moreover, these learning algorithms will require a huge computational effort during training. Therefore, it is reasonable to develop a CVNNs and a fast learning algorithm to overcome the above-mentioned issues. A learning in complex domain may require the input, output, weights, and activation functions all in complex domain. The second approach is more involved in the sense that there is a need to define a complex activation function, complex error function (EF) and based on this definition a new learning algorithm has to be introduced in complex domain. The benefit of this approach is that it yields a more efficient structure than from real domain neuron both in terms of computational complexity as well as performance. Moreover, this approach is realistic because it take care of phase information along with independent components during learning process.

The desired properties of a CVNN were stressed by the fully complex-valued multi-layer perceptron (CMLP) network and its gradient descent-based learning algorithm was derived by Kim and Adali [11]. Subsequently, a fully complex-valued radial basis function (CRBF) network and its gradient descent-based learning algorithm was developed in [6, 12]. Further studies have shown that the orthogonal decision boundaries of a CVNN with a split-type activation function provide them with superior decision making ability than their real-valued counterparts [2]. Its 2D structure of error propagation reduces the problem of saturation in learning and offers faster convergence. This also generated an increased interest among researchers to develop complex-valued classifiers to solve real-valued classification problems. The multi valued neural network (MVNN) [3], the single layer network with phase encoded inputs [4, 13], complex-valued extreme learning machine (CELM) [14] and the phase encoded complex-valued extreme learning machine (PE-CELM) [5] are some of the CVNNs available in the literature. CVNN is a generalization of the ANN to the complex domain where all parameters and functions are in complex domain. A recent investigation (Tohru Nitta 1997 and B K Tripathi 2011) reports that the size of the CVNN could be smaller than that of an ANN for the same problem (as each complex variable can take two real variables. Also the weights are complex which implies that they hold twice the information as real weights would) but it yields better and accurate results in comparison to equivalent real-valued neural network (RVNN). The application field of complex domain neuron is very wide. It is not easy to imagine areas dealing with 2-D parameters without the realm of complex numbers.

3.1.2 Out Performance Over Real Domain Neuron

In [7], author described a set of complex-valued learning patterns using two rules. This pattern set has a clear correspondence with popular XOR problem in real domain, therefore later named as ***C*XOR** problem. It has been used to test the learning and classification ability of real and complex domain neuron (with sigmoidal function) through corresponding learning algorithms. The ***C*XOR** problem is one of the best nonlinearly separable pattern associator in complex domain like popular XOR problem in real domain. The pattern set in ***C*XOR** is defined by following rules:

- the real part of output is 1 if first input is equal to second input, otherwise it is 0.
- the imaginary part of output is 1 if second input is either 1 or j, else it is 0.

This test problem is chosen with an eye toward testing the ability of the real and complex domain neuron. The comparative learning speed is presented in Fig. 3.1 and some of the observations on the key issues of optimization are given in Table 3.1. The neural network based on complex domain neuron is comparatively faster than real domain neuron while the space complexity (number of learning parameters: synaptic weights and bias) is only half whereas the time complexity (number of computations per learning cycle: addition, subtraction, division, and multiplication) remained almost same.

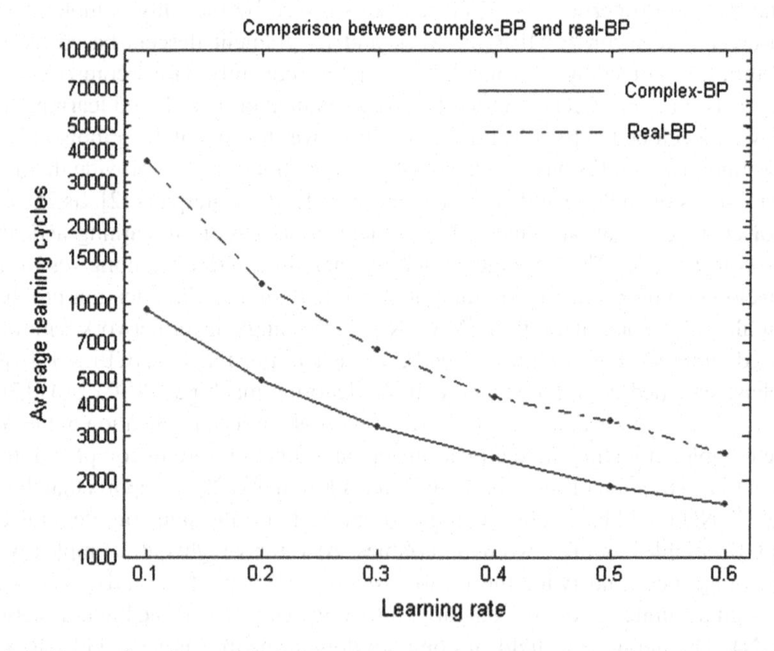

Fig. 3.1 Average learning speed of a sigmoidal neuron in real and complex domain for CXOR problem

Table 3.1 Complexity of the neural network with complex domain neuron and real domain neuron for *C*XOR Problem

Network	Time complexity			Space complexity		
Based on	× and ÷	+ and −	Sum	Weights	Thresholds	Sum
Complex domain neuron 2-4-1	134	92	226	24	10	34
Real domain neuron 4-9-2	150	76	226	54	11	65

3.2 Activation Functions in Complex Domain

There were a spectrum of functions that performed activation to a real variable-based neuron of the ANN. In fact, pieces of well-known curves could be joined to tailor curves that looked like the sigmoid (and possessed derivatives to the required order) of the ANN, which when applied as activation functions did obtain a convergent sequence of error with epochs. On the other hand heresy to the flexibility in the choice of activation functions in the ANN, the CVNN is constrained by some additional facts that are embedded into the complex domain in which it operates. The Liouville Theorem [15, 16] imposes more constraint on the fully complex valued functions. The constraint imposed by the theorem is additional in the complex variable setting as an equivalent did not exist in the real variable based ANN. Apart form fully complex-valued activation functions, a wide variety of activation functions have been investigated including the split-type function [9, 17, 18], phase-preserving function [7] and circular-type function [19].

Complex Activation Functions (*C*AF) and their characteristics depart from the traditional activation functions of the ANN in real domain. First, these functions have real and imaginary parts each of which individually are functions of two variables that make them surfaces in three-dimensional space, the analyticity of which plays a vital role of course. Second, there exist additional constraints imposed by the complex plane in the form of Liouville Theorem (Ahlfors 1979) that restrict the choice of functions that could be used as the *C*AF. As a result, tailoring new *C*AF by sewing pieces of surfaces along the common boundary would not be an acceptable proposition as the analyticity of the function developed this way should be established at each point on the boundary and later the construct must be verified to have cleared the constraint imposed by the Liouville Theorem. This restriction is unlike real value activation functions that could be easily tailored by joining differentiable functions and establishing differentiability at finitely many points (where each piece joins up with the next). In this chapter, two main broad directions of *C*AFs reported by researchers have been investigated in the above context.

Theorem 3.1 *Liouville Theorem states that 'If a complex valued function is both analytic and bounded through out the complex plane, then it must be a constant*

function.' Hence, for a nontrivial complex-valued function, analyticity and boundedness cannot hold together.

The contra-positive of the statement of theorem puts it in the most usable form as it spells out conditions that serve as search tools when one embarks on a search for ***C***AF. It states that a nonconstant complex valued function should either be nonanalytic and bounded or analytic and unbounded or nonanalytic and unbounded. The three possibilities must be verified, as the new complex activation must clear this constraint. It hence follows that at least one of the above three conditions must be satisfied for otherwise the activation would turn out trivial, the constant complex function.

3.2.1 Why Vary Activation Functions

A practical implementation of learning is not easy task in CVNN as in case of real-valued counter part. It depends on several factors but the important are the chosen architecture of CVNN and activation function of complex-valued neurons. Literature review in the area revealed that many questions about the architecture of the CVNN and activation functions employed have remained open as investigators have either not addressed them or have given partial information on these points of interest. It was discovered during the course of research that some reported results contradicted each other. For example, Leung and Haykin (1991) claimed that the fully complex activation function (***C***AF) given by the formula, Eq. 3.1 (where z is the net potential to the neuron in complex domain), converged in their experiment, while T Nitta (1997) reported that the same ***C***AF never converged in his experiments. Leung and Haykin (1991) also stated that the singular points of the ***C***AF could be circumvented by scaling the inputs to a region on the complex-plane but no procedure to implement this was described. The above facts clearly indicate the need for comprehensive investigations to establish the properties of the CVNN. It is very important to keep in mind that the basic complex theory states to choose either analyticity or boundedness for an activation function of complex-valued neuron, in view of Liouville's theorem.

It is noteworthy that there is a wide direction in complex-valued neurons depending on different activation functions and architecture; hence there are different complex-valued neural networks. The first complex version of steepest descent learning method made its appearance when Widrow et al. (1975) in USA presented the complex least mean square (LMS) algorithm [20]. Researches in this area took a significant turn in early 1990s, when various scientists have independently presented complex back-propagation algorithm (BP) with different activation functions. In 1991, Haykin considered complex-valued neuron on the basis of straightforward extension of real sigmoid activation function for complex variables [21]. Hence evolved fully complex-valued neuron based on analytic property of function in complex domain. They analyzed the dynamics with partial derivatives in real and imaginary parts. Later, in 2002, T Adeli et al. presented fully CVNNs with different analytic activation functions [11]. In 1992, Piazza considered a complex-valued neuron on the basis of 2D extension (real-imaginary type or split-type activation function) of real sigmoid function [17]. Hence evolved a complex-valued neuron based on boundedness property of function in complex domain in place of analyticity. Later T Nitta (1997) and B K Tripathi (2010) have formally compiled the 2D transformation property and wide acceptability in variety of tasks, dealing with split-type function [7, 18, 22]. They confirmed the ability of this function to handle magnitude-phase relationship properly during learning. It is also important to mention here that the development of CVNN is becoming more and more popular not only in complex-valued problem but also in real-valued problem [3–6, 8].

This generated an increased interest among researchers to develop many other neurons based on different architecture and function of activation over the field of complex domain. In 2007, Aizenberg formally presented the multi valued neuron (MVN) [3] using a multiple-valued threshold logic to map the complex-valued inputs to discrete outputs using a piecewise continuous activation function, which maps the complex plane into the unit circle. The MVN learning is reduced to progress along the unit circle which is based on a straightforward linear error correction rule and does not involve a derivative of activation function. In 2009, Murase developed complex-valued neuron with phase encoded inputs [4, 13] specially for real-valued problems. He obtained complex-valued features by phase encoding the real-valued features between $[0; \pi]$ using the transformation $z_t = e^{(i\pi x_t)}$; where x_t are the real-valued input features normalized in [0, 1]. He proposed [4, 13] two CAF that map complex values to real value outputs, by dividing the net potential (weighted summation) of neuron into multiple regions for identifying the classes (like real-valued neurons). Both the functions are differentiable with respect to real and imaginary parts of the net potential which in turn makes it possible to derive the gradient based learning rules. In contrast to single real-valued sigmoid neuron which saturates only in two regions and can only solve linearly separable problem; the single complex-valued neurons developed by Murase saturates in four regions, hence significantly improves their classification capability. This idea of phase encoding, to transform the real-valued input features to the Complex domain, is extended by Sudararajan to develop PE-CELM (2011) [5, 14] and fully CRBF classifier (2012) [6] for real-valued classification problems. The phase encoded transformation maps the real-valued input features into the first and second quadrants of the complex plane, completely ignoring the other two quadrants. Therefore, the transformation does not completely exploit the advantages of the orthogonal decision boundaries. In 2013, Suresh circumvented this issue by employing circular transformation to map the real-valued input features to the complex domain, in fully complex-valued relaxation neural network (FCRN). The circular transformation effectively performs a one-to-one mapping of the real-valued input features to all four quadrants of the complex domain. Hence, it efficiently exploits the orthogonal decision boundaries of the FCRN classifier. FCRN is a single hidden layer network, with a Gaussian-like hyperbolic secant function (sech) in the hidden layer and an exponential function in the output layer for neuron's activation. He claimed to approximate the desired output more accurately with a lower computational effort. Majority of researchers in complex domain have accepted that the complex-valued neuron with a split-type CAF is very easy to implement, provide superior decision making ability and effectively applicable in all problems in complex domain.

3.2.2 Properties of Suitable Complex-Valued Activation Function

The performance or approximation offered by neural networks depends to a great extent on its activation function. Therefore, the study of activation functions and their derivative is important for its proper choice. In case of RVNN, the chosen activation function should be a smooth (continuously differentiable) and bounded function; such as sigmoidal function. When a real domain is extended to the complex domain, there exist certain difficulties involved in the appropriate choice of an activation function due to the set of properties that a suitable complex activation function must possess. Let a complex-valued function be

$$f(z) = f_{\Re}(z_{\Re}, z_{\Im}) + j\, f_{\Im}(z_{\Re}, z_{\Im}) \quad \text{where} \quad z = z_{\Re} + z_{\Im}$$

A suitable activation function is supposed to possess the following reasonable properties:

- $f(z)$ must be nonlinear in ($z_{\Re}$ and $z_{\Im}$). If $f(z)$ is linear:
  - There is no advantage in using learning algorithm.
  - The capabilities of the network are severely limited if $f(z)$ is linear.
- $f(z)$ is bounded.
  - This is true if and only if both $f_{\Re}(z_{\Re}, z_{\Im})$ and $f_{\Im}(z_{\Re}, z_{\Im})$ are bounded.
  - Both of these are used during training and even if one of them is unbounded, it can result in divergence.
- $f(z)$ is such that error $e \neq 0$ and inputs $z \neq 0$ implies $\nabla W(E) \neq 0$
  - Noncompliance with this condition is undesirable since it would mean that even in the presence of a *nonzero input* and *nonzero error*, it is still possible that no learning takes place ($\nabla W(E) = 0$).
  - Both of these are used during training and even if one of them is unbounded, it can result in divergence.
- The partial derivatives $f'_{\Re}(z_{\Re})$, $f'_{\Im}(z_{\Im})$, $f'_{\Re}(z_{\Im})$ and $f'_{\Im}(z_{\Re})$ exist and are bounded.
  - These partial derivatives are used during training and the weights are updated by amounts proportional to these partial derivatives, therefore they must be bounded.
- $f(z)$ is not entire.
  - According to Liouville's Theorem if $f(z)$ is entire and bounded on the complex palne, then $f(z)$ is a constant function.

3.2.3 Complex Activation Functions

The function of activation in the neuron of the CVNN is a complex-valued function unlike the RVNN where the functions was real-valued. The activation function for CVNN ($\mathbb{C}$AF) is an extension of what existed as the function of activation in RVNN but need improvisation to suit to complex variable based ambience. In view of theory of complex variable and complex-valued function various researchers have given different extension or formulation of $\mathbb{C}$AF. A comprehensive investigation of the complex domain error backpropagation learning had been tried with different activation functions. This section presents two prominent and popular approaches of defining complex-valued activation functions with their general properties resulted on variety of applications. The complex function unlike real function is in two dimension because it is two variable function (real and imaginary parts), hence surface of activation function is in three dimensional space as both real and imaginary parts of the complex function are function of real and imaginary parts of variables.

3.2.3.1 Simple Extension of Real Activation Function

It is well known that a typical and most frequently employed activation function in the real domain has a sigmoidal behavior. Leung and Haykin [21], Kim and Guest [23] have proposed a straightforward extension of this function to the complex domain as:

$$fc(z) = \frac{1}{1+e^{-z}} \tag{3.1}$$

where $z = x + j\,y$ in the problem is the net input (weighted inputs) at the fan-in to each neuron. This is referred to as Haykin activation function (in honor of its discoverer) and is indeed analytic. Similarly, other commonly used activation functions e.g. $\tanh(z)$ and $\exp^{-z^2}$ are extended from real to complex domain. Though the formulation of the complex-valued neuron with an analytic complex functions, as given in Eq. 3.1, seemed to be natural and to produce many interesting results, however there was problem of unboundedness for some of the input values. Haykin et al (1991) have proposed that this problem can be avoided by scaling the input data to some region on the complex plane. The plots of activation function are shown in Fig. 3.2. The properties of analyticity and unboundedness can be appreciated from the plots.

After some algebra, the real and imaginary parts of the Haykin activation function (Eq. 3.1) are respectively:

$$\frac{1+e^{-x}\cos(y)}{1+e^{-2x}+2e^{-x}\cos(y)} \qquad\qquad \frac{e^{-x}\sin(y)}{1+e^{-2x}+2e^{-x}\cos(y)} \tag{3.2}$$

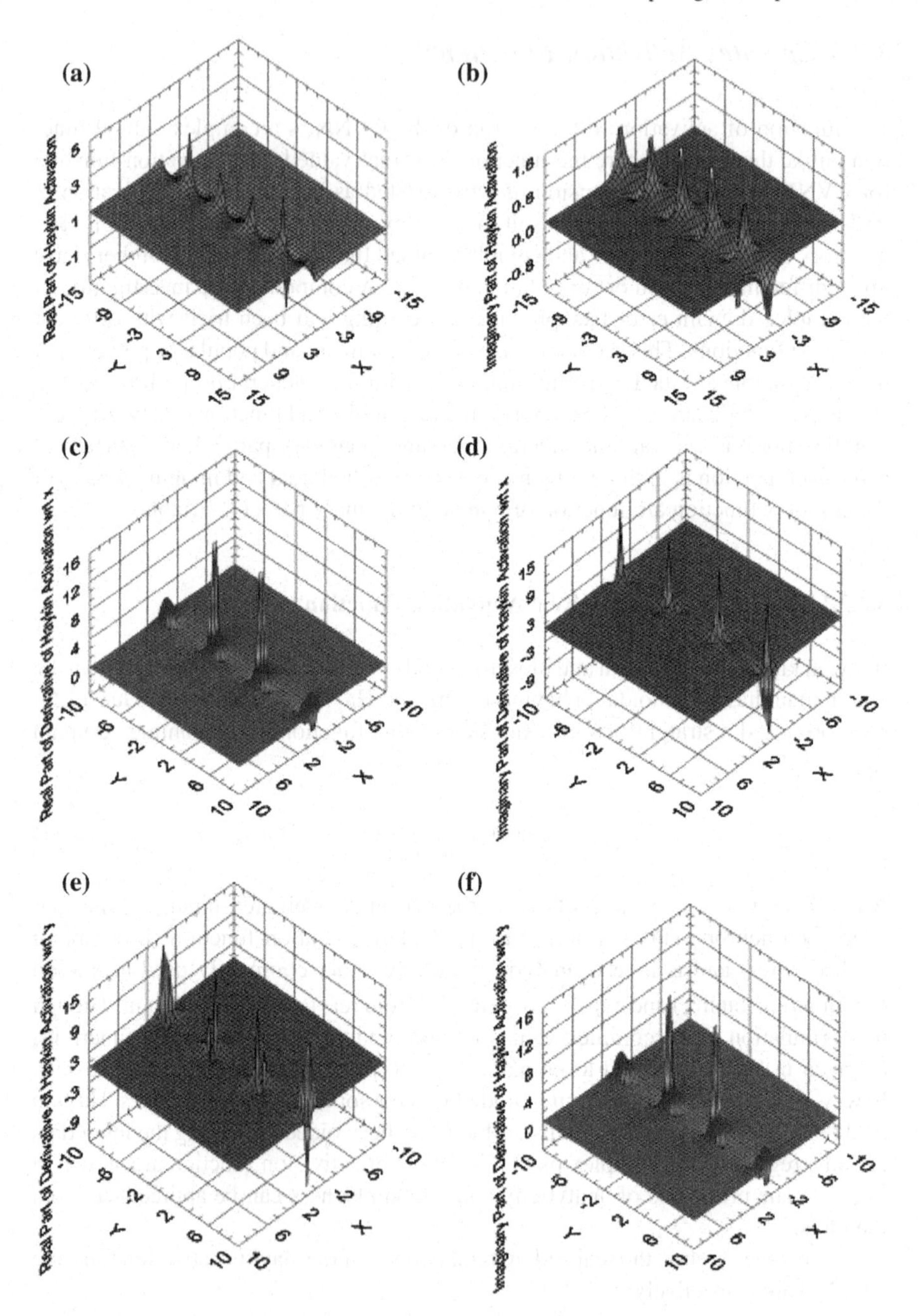

Fig. 3.2 **a** Real part of Haykin activation (sigmoid function). **b** Imaginary part of Haykin activation. **c** Real part of the derivative with respect to x. **d** Imaginary part of the derivative with respect to x. **e** Real part of the derivative with respect to y. **f** Imaginary part of the derivative with respect to y

It should be noted that on equating the denominator of both expressions to zero, we observe that the function goes unbounded at points of the type $(0, (2n + 1)\pi)$ for any natural number 'n'. Figure 3.2a is a plot of the real part and Fig. 3.2b is a plot of the imaginary part of Haykin activation function. Both figures are characterized by prominent peaks (singular points). Invoking the Liouville theorem, we find that the functions are unbounded and hence qualify as activation functions. To avoid the singular points, the inputs to the neuron should be scaled to a region that is devoid of them [21]. In the update rule for the weights with Haykin activation function the derivative of Eq. 3.1 with respect to z needs to be computed, which is $f_C(z)$ $(1 - f_C(z))$. The surface plots of the expressions for real and imaginary parts of the derivative of Haykin's activation function with respect to real part x the imaginary part y[1] are displayed in Fig. 3.2c–f, respectively. All surfaces, as can be seen, are characterized by peaks. The backpropagation learning algorithm developed with the Haykin activation function has singularities at the countably many points, the derivative of the Haykin activation also vanishes at these points.

3.2.3.2 Problem with Haykin's Activation Function

When the domain of conventional activation like sigmoid in Eq. 3.1, $\tan(z)$ or $\exp^{-z^2}$ is extended from real to complex, it is seen that if z approaches any value in the set $\{0 \pm j\,(2n + 1)\,\pi\}$ where n is integer, then $|f_C(z)| \rightarrow \infty$ thus $f_C(z)$ is unbounded. It was also suggested in Leung and Haykin [21] that to avoid the problem of singularities in the sigmoid function $f(z)$, the input data should be scaled to some region in the complex plane. The position of singularities disturb the training scheme as whenever some intermediate weights fall in the vicinity of the singular points, it was observed that the whole training process down the line receives a jolt. This is revealed by the error plot of the function, which is characterized by peaks. The typical point scatter shown in the Fig. 3.3a is a distribution of the hidden layer weights as the training process is on. The figure shows four singular points of the Haykin activation that are completely engulfed by the cloud of points. As can be observed, they cluster around the some singular points which eventually results in the peak type error-epochs characteristic. The typical error function graph with Haykin activation is shown in Fig. 3.3b. The training process produced many peaks as a result of the singular activation configurations encountered because of the activation function's singular points. The complex backpropagation developed over this activation functions fails to solve many problems.

In case of complex-valued networks, T Adali (2003) broadly categorized the fully complex-valued activation functions by their properties [11] into three types. It was also shown that universal approximation can be achieved for each of them. The first type of complex-valued functions concerns the functions without any singular points. These functions can be used as activation functions and the networks with this type of activation functions are shown as good approximators. Although some of

---

[1] Finding the expressions for real and imaginary parts of the derivative of Haykin's activation function with respect to real part x the imaginary part y are left for interested readers.

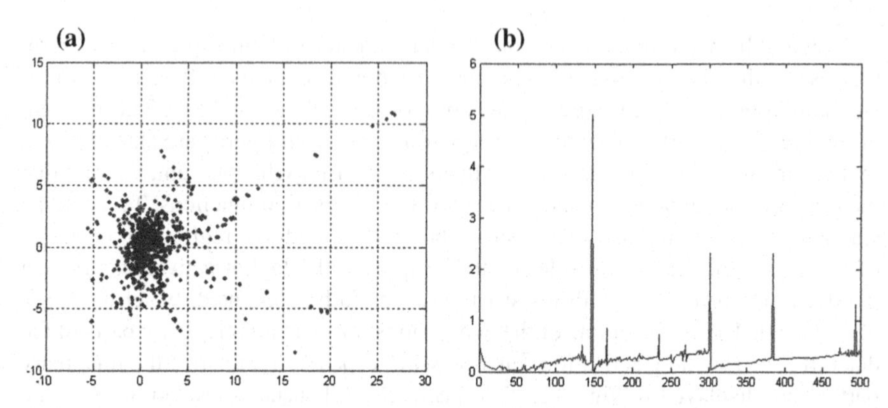

Fig. 3.3 **a** *Encircled* are the points where Haykin activation and its derivative vanish. Four singular points of the Haykin Activation Function Eq. 3.1 (explicitly, (0,1), (0,−1), (0,3), (0,−3)) have been fully engulfed by the point scatter. The points (0,3) and (0,−3) are shown *encircled* while the other two points are completely inside the cloud of points. **b** Typical convergence pattern with Haykin activation, shows the peaks formed during a run of training in error-epochs characteristic

the functions are not bounded, they can be used by introducing bounding operation for their regions. The second type concerns the functions having the bounded singular points, e.g., the discontinuous functions. These singularities can be removed and thus they can also be used for activation functions and can achieve their universality. The last type is for the functions with the so-called essential singularities, i.e., their singularities cannot be removed. These functions can also be used as activation functions, with the consideration of restricting the regions for them so that their regions never cover their singularities. But, backpropagation is a week optimization procedure, therefore there exist no mechanism of constraining the values that the weights can assume. Thus, the value of z that depends upon both the inputs and the weights, therefore they can take any value on the complex plane. Based on this observation. the suggested remedy is inadequate and there is a need to find some other activation functions, which can satisfy the resonable properties of suitable activation function.

3.2.3.3 Two-Dimension Extension of Real Activation Function (Split-Type Function)

By Liouville's theorem, an analytic function can not be bounded on all of the complex plane unless it is a constant, therefore the analyticity of the complex function can not be preserved because boundedness is required, to ensure convergence. Benvenuto and Piazza [17] have suggested an alternative complex activation function to avoid the conflict between the requirements of an activation function and Liouville's theorem, in general. The characteristic function f_C appears in function of real and imaginary parts of weighted inputs separately (a split-type activation function) in a multilayer neural network framework:

$$f_C(z) = f(x) + j\, f(y) \tag{3.3}$$

where $z = x + j\,y$ is a complex variable and $f \in \boldsymbol{R}$ is conventional real-valued activation function as given below:

$$f(x) = \frac{1}{1 + e^{-x}} \tag{3.4}$$

In split activation function, a nonlinear function is applied separately to the real and imaginary parts of the net potential (aggregation at the input) of the neuron. Here sigmoidal activation function (Eq. 3.4) is used separately for real and imaginary part. This arrangement ensures that the magnitude of real and imaginary part of $f_C(z)$ is bounded between 0 and 1. But now the function $f_C(z)$ is no longer analytic (holomorphic), because the Cauchy-Riemann equation does not hold:

$$\frac{\partial f_C(z)}{\partial x} + j\frac{\partial f_C(z)}{\partial y} = \{1 - f(x)\}\, f(x) + j\,\{1 - f(y)\}\, f(y) \neq 0$$

The 2D extension of the real sigmoid function, as in Eqs. 3.3 and 3.4, always yield a bounded behavior of activation function $f_C(z)$. So, effectively the holomorphy is compromised for the boundedness of the activation function. Figure 3.4 shows the output magnitude of the neuron with respect to the real and imaginary parts of the input for function definition given in Eq. 3.3.

It has been proved that CVNN with activation function, Eq. 3.3, can approximate any continuous complex-valued function where as the one with a analytic activation function, Eq. 3.1, proposed by Kim and Guest in 1990, Haykin in 1991 and tanh(z) by Adali in 2000 can not approximate any nonanalytic complex-valued function [2]. Thus, CVNN with activation function, Eq. 3.3, appeared as a universal approximator in all researches. In 1997, Nitta [7] and later Tripathi [8, 18, 24] has also confirmed the stability of learning of CVNN with this activation function through

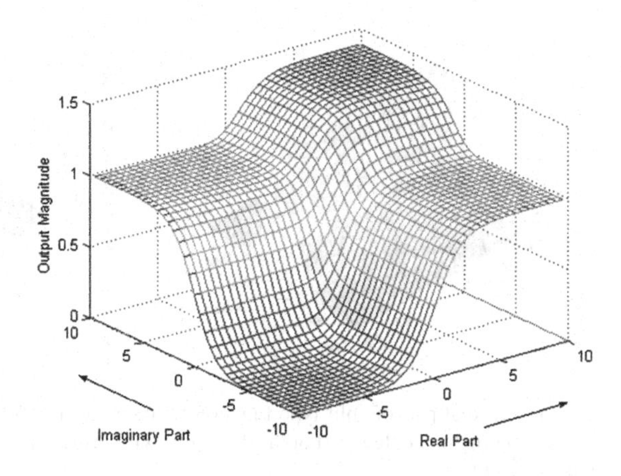

Fig. 3.4 Output magnitude of the split-type activation function

wide spectrum of computer simulations. Their simulations for conformal mappings (geometric transformations) were not possible without the ability to process signals maintaining magnitude and phase relationship properly in activation function, Eq. 3.3, explained in Chap. 5. The accurate estimation of both magnitude and phase of the complex-valued signals are important in many real-world applications. Thus, learning in CVNN with activation function, Eq. 3.3, ensures the flow of complex-valued signals preserving the magnitude and phase relationship.

The real part of activation function defined with equation Eqs. 3.3 and 3.4 is independent of y and its imaginary part is independent of x. The derivative of the activation function with respect to the real part of the argument, x, is free from the imaginary part. Similarly, the derivative of the activation function with respect to the imaginary part of the argument, y, is pure imaginary. The surface plots of the real and imaginary parts of the function and their derivatives are shown in Fig. 3.5. It can be easily seen that the function does not have any singular points and nor

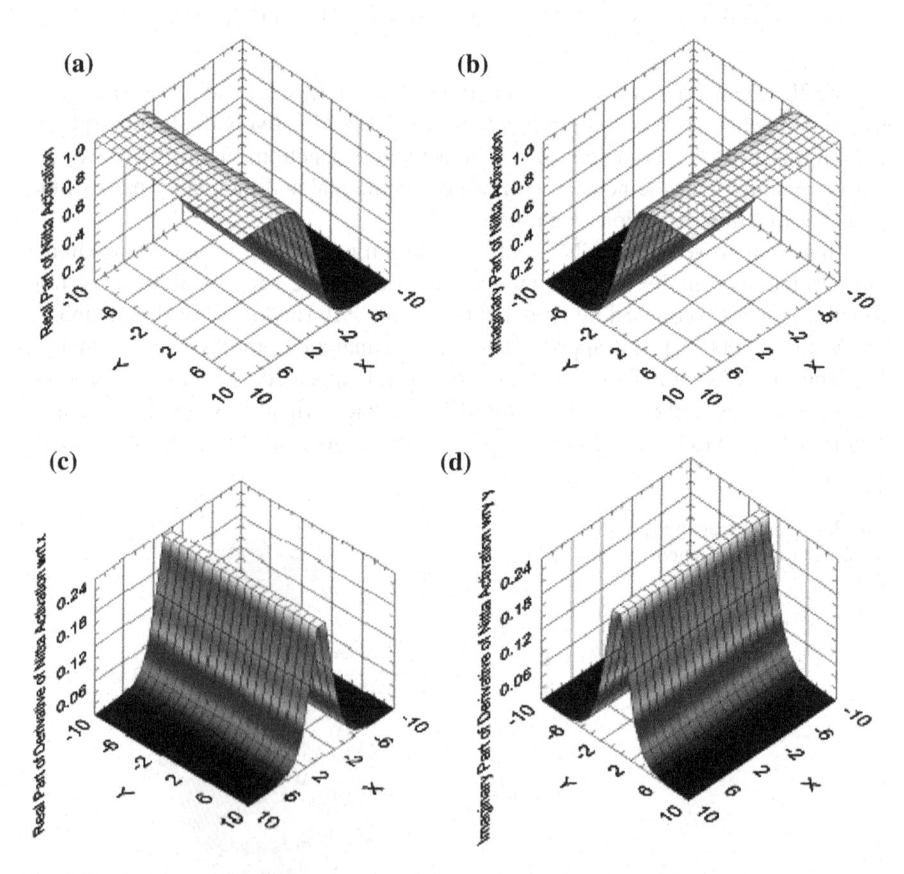

Fig. 3.5 **a** Real part of split-type function (Nitta activation function), **b** Imaginary part of Nitta activation function, **c** Real part of the derivative with respect to x, **d** Imaginary part of the derivative with respect to y

do the derivatives possess any singular points. Recent studies have shown that the orthogonal decision boundaries of a CVNN with a split-type activation function provide them superior decision-making ability than their real-valued counterparts [2, 9]. The learning rules in complex backpropagation (***C***BP) algorithm for CVNN with this function (Eq. 3.3) posses linear combination of derivatives of real and imaginary components of an output function; hence effectively reduces the standstill in learning. Such CVNN has outperformed not only in complex-valued problems but also in real-valued problems, Chap. 7.

3.3 Error Functions in Complex Domain

The role of error function for data analysis is emphasized from a CVNN viewpoint in this section. In almost all investigations and applications involving ANNs, the BP applied is the one developed over a quadratic EF. This error function (EF) may not perform satisfactorily for all real life data with function approximation and classification. Literature points out many other functions that can take the place of the traditional quadratic EF in data analysis but these EF-based applications have been studied from a statistics viewpoint and not from a neural network viewpoint. Incidentally, Rey (1983) pointed out that in statistical analysis replacing EF can yield better results. It must be stressed here that the EF over, which the complex-valued back-propagation algorithm (***C***BP) was built recently also is the quadratic EF. However, a few researchers (Werbos and Titus (1978), Gill and Wright (1981), Fernandez (1991), Ooyen and Nienhaus (1992)) have used different EF with their BP. While complex error funtions (EFs) have not been investigated in a systematic fashion. Replacing the EF assumes importance because practical data are prone to measurement errors and outliers. If the quadratic EF were retained for analyzing data prone to outliers and other errors, the curve of best-fit would not be appropriate because the cost that accrues to the chosen EF would get enhanced due to the power term for far-off points (outliers). Instead, if the quadratic EF were replaced with an absolute function for example, the curve-fitting scheme for noise-and-error-prone data would be more evenly placed because the cost accrued due to the these data would be of the same order as the actual data points. This even-weighting results in a better curve-fit than the one obtained by quadratic error based approximation. This section surveys some EFs and studies BP and the ***C***BP from an EF viewpoint.

3.3.1 Why Vary Error Functions

Most of the statistical methods used for practical data analysis employ a Quadratic EF. Further, almost all learning algorithms for ANN reported in the literature, use a mean square error deviation between the actual and predicted output as the EF. Rey (1983) pointed out that by varying the EF in an optimization scheme, the result can be improved substantially. The statement was substantiated by demonstrating that an Absolute EF-based optimization solved a curve-fitting problem more efficiently than the standard Quadratic EF-based optimization. The observation was attributed to the fact that the cost accrued is greater if a power term exists; the outlier points hence distort the optimization scheme. On the other hand, an Absolute EF is equi-poised toward data points and the outliers and therefore the ill-effects of the outliers are more balanced in the scheme resulting in a better curve-fit.

Earlier Werbos and Titus (1978), Gill and Wright (1981) also discussed the idea of changing the EF in an optimization scheme. Fernandez (1991) implemented some new EFs that were designed to counter the ill-effects local minima by weighting the errors according to their magnitudes. Matsuoka (1991) reported BPA based on Logarithmic EF and elimination of local minima. Ooyen and Nienhaus (1992) used an entropy type EF and showed that it performs better than the Quadratic EF-based BP for function approximation problems. It is worth to mention here that the data set obtained while practically experimenting is prone to system noise, process noise, and measurement errors (like parallax). The outlier points contribute to the offset in the solution to the curve-fitting problem (Rey 1983). There exist two approaches to tackle the undesirable affects of spurious data points. The first approach demands that these points be eliminated completely (by some data processing technique) and later after weeding out these points, subject the data to a Quadratic EF-based optimization scheme and obtain a solution, which can be termed ideal approach. The second approach, as explained in Rey (1983) requires incorporating a modified EF that would by the nature of design and construction have useful properties to bypass the ill-effects of the spurious points in the data and obtain a better fit of curve to the data set than the Quadratic EF. Therefore, there is a requirement to identify the properties that useful EFs in real and complex domain must satisfy and subsequently investigate learning rules so that user can overcome from the drawbacks of different functions and select for better optimization. The next section presents various important EFs in real and complex domain with viewpoint of backpropagation learning algorithm in corresponding real (BP) and complex (**C**BP) domain.

3.3.2 Definitions and Plots of Error Functions

This section presents few prominent EFs in view of neural network learning procedure, though these EFs were proposed in literature to implement an m-estimators approach for bypassing or reducing the ill-effects of the outliers. The functions in each case have been generalized in a way as to retain the form and yet be operational in the complex variable setting. This also makes sure that the surface plot of the function is close to the plane plot of the same. The definition EFs in real and complex domain and their corresponding plane and surface graphs displaying their overall shape are demonstrated. It could be appreciated from the graphs that retaining the form of the function manifests itself in the plots as they resemble their counterparts on the plane. Let $e \in \boldsymbol{R}$ and $\varepsilon \in \boldsymbol{C}$ be the error and E be the EF for optimization in artificial neural network. The learning algorithms in complex domain derived from any EF ($E : \boldsymbol{C} \longrightarrow \boldsymbol{R}$) explicitly minimize only the magnitude error.

3.3.2.1 Quadratic Error Function

The quadratic EF ($E : \boldsymbol{R} \longrightarrow \boldsymbol{R}$) for practical data analysis in real domain is developed by mean squared error, which is given by

$$E = \sum_{n} e_i^2 \tag{3.5}$$

The complex quadratic EF ($E : \boldsymbol{C} \longrightarrow \boldsymbol{R}$) is defined to be

$$E = \sum_{n} \varepsilon_i \varepsilon_i^* \tag{3.6}$$

where n is the number of outputs and superscript $*$ represents the complex conjugate of variable. This is widely accepted as the standard EF (Bose and Liang 1996). Its plane and surface plots are given in Fig. 3.6a and b, respectively.

3.3.2.2 Absolute Error Function

Absolute error is one of several robust functions that displays less skewing of error due to outliers. A small number of outliers are less likely to affect the total error and so they do not affect the learning algorithm as severely as the quadratic error.

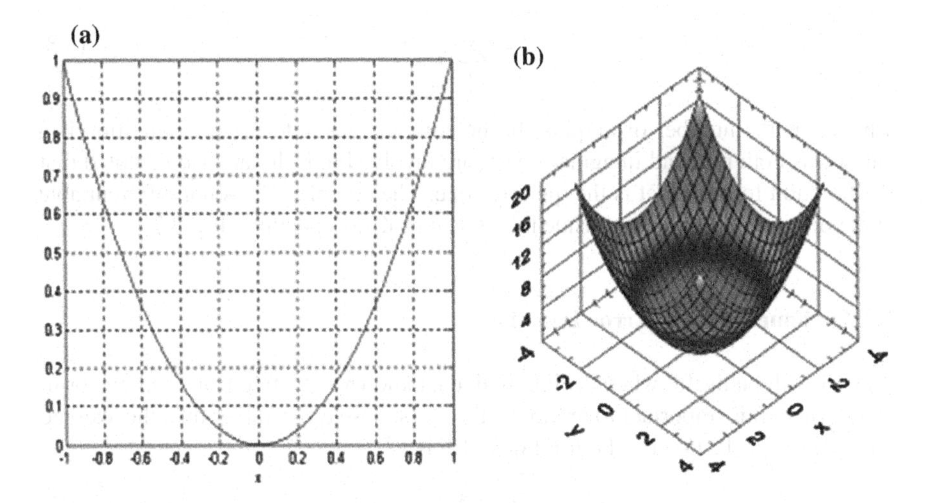

Fig. 3.6 **a** Plane and **b** Surface plots for quadratic EF

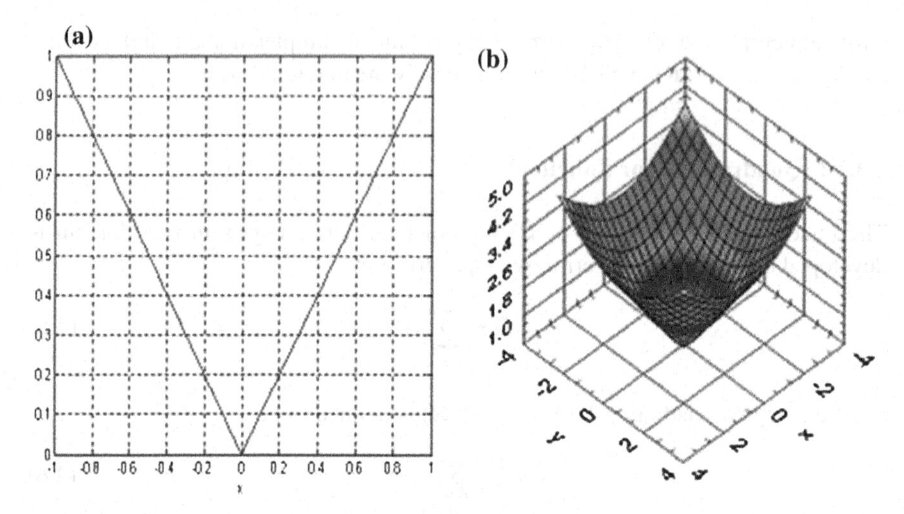

Fig. 3.7 **a** Plane and **b** Surface plots for absolute error function

When it is applied to data prone to outliers, analysis obtains a better curve of fit than quadratic EF. Absolute EF is a generalisation of the absolute function defined to compute errors using the absolute function yet retaining the functional form. The absolute error in real domain is given by (Fig. 3.7)

$$E = \sum_{n} |e_i| \tag{3.7}$$

The complex absolute EF ($E : \boldsymbol{C} \longrightarrow \boldsymbol{R}$) is defined to be

$$E = \sum_{n} \sqrt{\varepsilon_i \varepsilon_i^*} \tag{3.8}$$

where n is the number of outputs. In the complex absolute EF, the definition is a surface as both real and imaginary parts are involved in it. It can also be noted that the function form in fact is the quadric cone. The complex EF is not differentiable at the origin as the function inside the radical is always positive (Fig. 3.7).

3.3.2.3 Fourth Power Error Function

This EF will be useful when dealing with data known to be free from outliers, or in cases where it is important to minimize the worst-case error, rather than the average error (Hassoun 1995). The Fourth Power EF is:

$$E = \sum_{n} e_i^4 \tag{3.9}$$

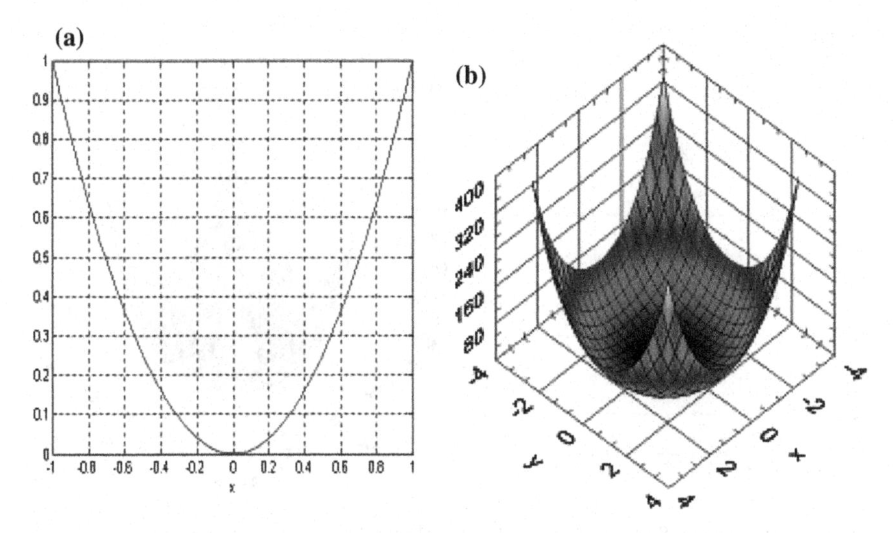

Fig. 3.8 **a** Plane and **b** Surface graphs of fourth power EF

The complex fourth power EF is defined to be

$$E = \sum_{n} \frac{1}{2} (\varepsilon_i \varepsilon_i^*)^2 \tag{3.10}$$

where n is the number of outputs. The plane and surface plots of fourth power EF are given in Fig. 3.8a and b respectively.

3.3.2.4 Minkowski Error Function

The Minkowski EF is given by

$$E = \sum_{n} |e_i|^r \tag{3.11}$$

The complex Minkowski EF is defined to be

$$E = \sum_{n} |\varepsilon_i|^r \tag{3.12}$$

where n is the number of outputs and $r \in \boldsymbol{R}$ is the Minkowski generalization parameter. The plane and surface plots of Minkowski EF are given in Fig. 3.9a and b respectively.

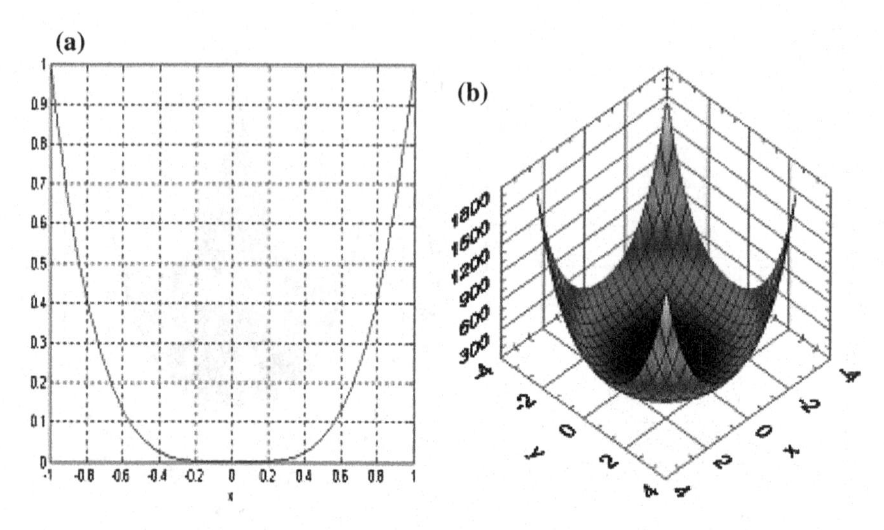

Fig. 3.9 **a** Plane and **b** Surface graphs for Minkowski EF for typical value of $r = 4$

3.3.2.5 Mean-Median Error Function

This EF takes the advantage of both the mean error function and the median EF. Hence, reduces the influence of large errors but at the same time retains its convexity. The mean-median EF is given by

$$E = \sum_{n} 2 \times \left(\sqrt{\left(1 + \frac{e_i^2}{2} \right)} - 1 \right) \tag{3.13}$$

The complex mean-median EF is defined to be

$$E = \sum_{n} 2 \times \left(\sqrt{\left(1 + \frac{\varepsilon_i \varepsilon_i^*}{2} \right)} - 1 \right) \tag{3.14}$$

where n is the number of outputs (Fig. 3.10).

3.3.2.6 Sine-Hyperbolic Error Function

This EF is steeper than the quadratic EF. Moreover the function is symmetric about the origin and hence the update involves two parts, the first is the gradient in the first quadrant while the second is gradient in the third quadrant. In both cases, the gradient is directed toward the origin. The sine-hyperbolic error function is given by

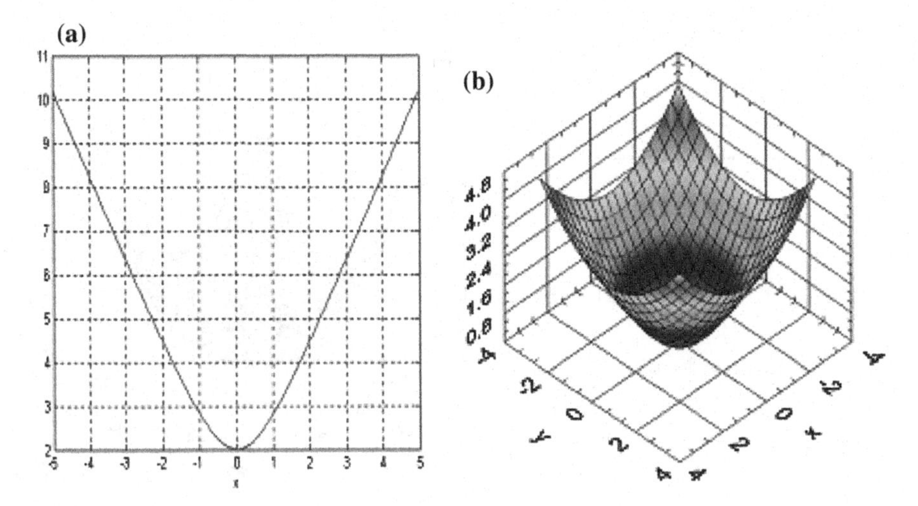

Fig. 3.10 **a** Plane and **b** Surface graphs for Mean-Median Error Function

$$E = \sum_{n} (\sinh(e_i)) \tag{3.15}$$

The complex sine-hyperbolic EF is defined to be

$$E = \sum_{n} \sinh\left(|\varepsilon_i|\right) \tag{3.16}$$

where n is the number of outputs (Fig. 3.11).

3.3.2.7 Cauchy Error Function

Cauchy EF, also known as the Lorentzian function, is one of the most robust functions of statistics. The tuning constant (c) for the EF is 2.3849. The cauchy's EF is given by

$$E = \sum_{n} \frac{c^2}{2} \ln\left(1 + \left(\frac{e_i}{c}\right)^2\right) \tag{3.17}$$

The complex Cauchy EF is defined to be

$$E = \sum_{n} \frac{c^2}{2} \ln\left(1 + \left(\frac{\varepsilon_i \varepsilon_i^*}{c^2}\right)\right) \tag{3.18}$$

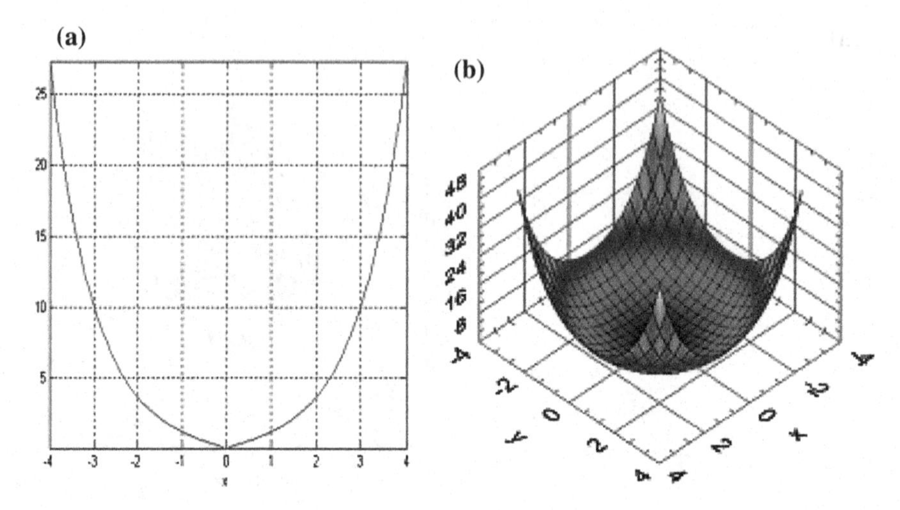

Fig. 3.11 **a** Plane and **b** Surface graphs for sinh error function

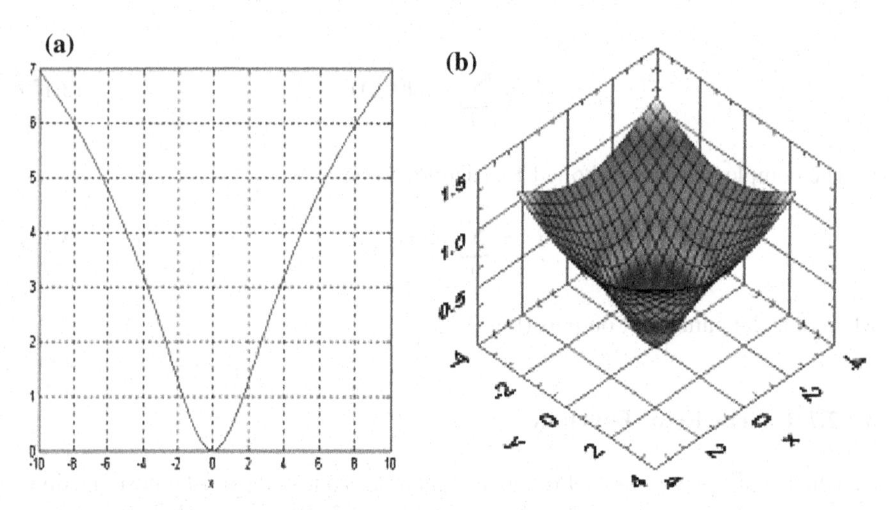

Fig. 3.12 **a** Plane and **b** Surface graphs for Cauchy error function

where n is the number of outputs and c is the tuning constant. The plane and surface plots of Cauchy EF are given in Fig. 3.12a and b respectively.

3.3.2.8 Huber Error Function

When dealing with noisy data, the training values may contain outliers with unusual deviation from the true underlying function. Huber function can be used to ignore these outliers, or at least reduce the ill effect they have on learning. The function

(Huber 1981) has good effects of quadratic and absolute EFs. The Huber EF is given by

$$E = \sum_n \frac{e_i^2}{2}; \quad \text{if } |e_i| < c \tag{3.19}$$

$$= \sum_n c\left(|e_i| - \frac{c}{2}\right); \quad \text{if } |e_i| \geq c \tag{3.20}$$

The complex Huber EF is defined to be

$$E = \sum_n \frac{\varepsilon_i \varepsilon_i^*}{2}; \quad \text{if } |\varepsilon_i| < c \tag{3.21}$$

$$= \sum_n c\left(|\varepsilon_i| - \frac{c}{2}\right); \quad \text{if } |\varepsilon_i| \geq c \tag{3.22}$$

where n is the number of outputs and c is the tuning constant. A typical value for c is 1.345. The Huber EF is defined piece-wise. The characteristic feature of the functions involve the quadratic error on the one hand and an absolute error on the other. The parameter, c in the definition is the point of demarcation to assign a domain of operation for each EF (Fig. 3.13).

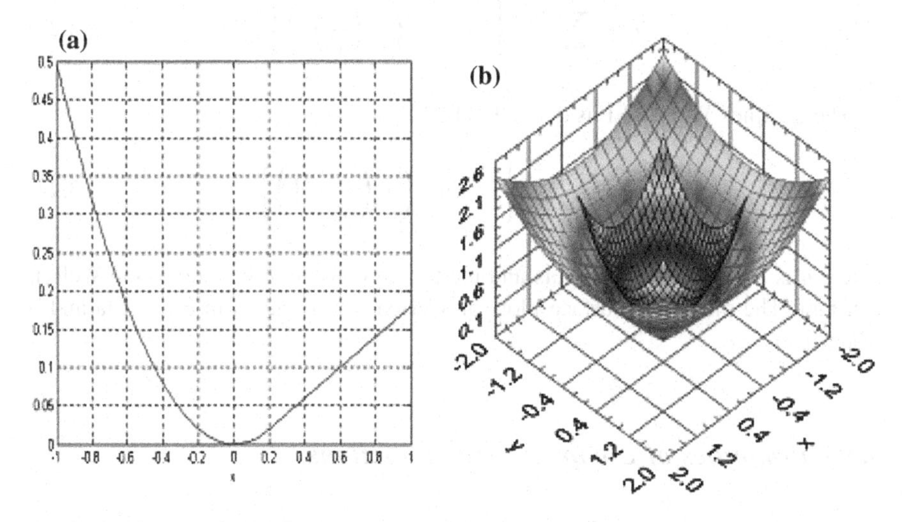

Fig. 3.13 **a** Plane and **b** Surface graphs for Huber error function

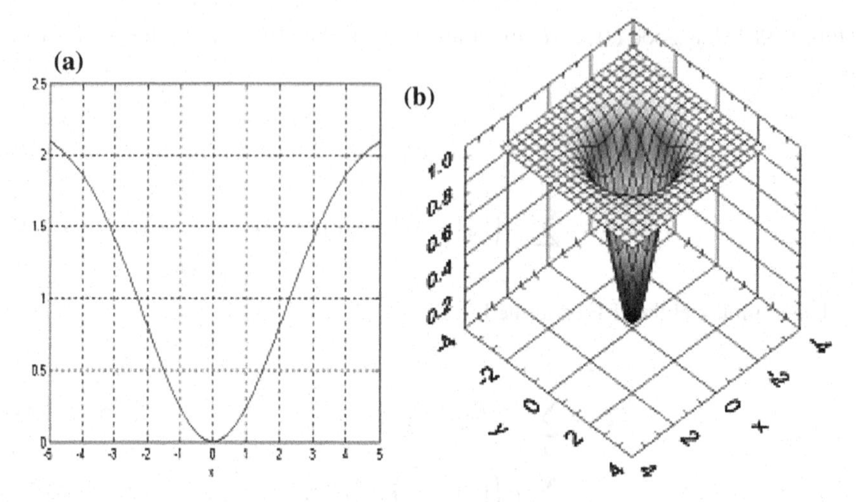

Fig. 3.14 **a** Plane and **b** Surface graphs Welsch error function

3.3.2.9 Welsch Error Function

This function reduces the influence of large errors. The typical value of the tuning constant is $c = 2.9846$. The Welsch error function is given by

$$E = \sum_{n} \frac{c^2}{2} \left\{ 1 - \exp\left(- \left(\frac{e_i}{c} \right)^2 \right) \right\} \tag{3.23}$$

The complex Welsch EF is defined to be

$$E = \sum_{n} \frac{c^2}{2} \left\{ 1 - \exp\left(- \left(\frac{\varepsilon_i \varepsilon_i^*}{c^2} \right) \right) \right\} \tag{3.24}$$

where n is the number of outputs and the tuning constant c was set to the Welsch constant. The plane and surface plots of Welsch EF are given in Fig. 3.14a and b respectively.

3.3.3 Properties of Complex Error Functions

Practical data is prone to outliers resulting from observational or experimental errors. If the quadratic EF were used to analyze data containing outlier points, the optimization scheme would result in estimates that depend significantly on the outliers as the quadratic EF assigns greater cost to far off points. A different error function, for example an absolute EF can perform better optimization than quadratic as the cost that accrues to the new EF due to outliers would be lesser than that with the quadratic EF resulting thereby in a better estimate. Similarly, other EFs possess advantageous situations which is contingent upon the quality of problem concerned.

The presented EFs which were generally employed in statistical analysis, will be useful for developing the real backpropagation (***R***BP) algorithm and the complex backpropagation (***C***BP) algorithm for training ANN and CVNN respectively. The update rule of the backpropagation algorithm demands the EF be at least once differentiable. Finitely many discontinuities or countably many of them can always be bypassed by defining the update rule accordingly by breaking the real line into finitely many or countably many intervals and developing a form of the update rule in each of the intervals separately. Of the many parameters that should be set for running the algorithm, the weights, biases, architecture, must be kept fixed to study the influence of the EF while EF-based training algorithms run. Each of the EFs has its unique properties that the statistical analysis cashes while implementing them. The EFs may be generalized to the complex variables in each case by retaining the form yet extended to accommodate complex numbers in backpropagation algorithm.

3.3.3.1 Absolute Error Function

The Absolute EF is continuous through out the real line and is differentiable at all points on the line except at the origin. As the real line is partitioned into two disconnected sets by the origin (the only point where the function is not differentiable), the update rule has a two-step definition—when the error is positive and when the error is negative. The absence of an index (the power, unlike the quadratic EF) is a distinguishing feature of this EF as this enables smoothing out the ill-effects of the outlier points that would otherwise have offset the best-fit of the optimisation scheme. The contribution to the EF from the outlier points would be on the same scale as the actual data points of the problem and hence the ill-effects due to spurious points are nullified to a great extent. On the other hand, if the data were normalised to a specific region so that all the entries in the data set are small real numbers lying in $[-1,1]$, the contribution from the outliers is once again on the same scale as the actual data points. The gradient for both parts in the definition in the update rule is directed toward the origin.

The complex EF as a whole is not differentiable at zero's as the function inside the radical is always positive. It has all the complex weights of the CVNN in its definition. The update rule for the CVNN steers the real part and the imaginary part of the weights to the minima separately. The problem of local minima that existed in the ANN repeats while studying the CVNN in general and this EF based algorithm in particular. As is clear the initial weight and the learning parameter decide how the training should progress. The dynamics of the real part depends not only on the real parts of the weights but also on the imaginary parts, as the updates of the real and imaginary parts are coupled (dependent on each other).

3.3.3.2 Fourth Power Error Function

The fourth power EF is smooth over the whole real line. Unlike quadratic, the function rises rapidly and is more convex with respect to the x-axis than the quadratic EF. The weight update is more rapid for error values greater than unity and the rate of training is diminished for fractional errors, lying in the interval $[0,1]$. The contribution from the cube term that results from the form of the EF is hence a parameter that enhances the update if the error is greater than unity and suppresses it if then error is fractional. The complex fourth power EF keeps the form of the real EF but is defined so as to be able to operate with complex errors. The EF unlike the complex quadratic function increases more rapidly for errors more than unity. The surface is rotationally symmetric about the z-axis. The surface is smooth for the derivatives of all orders exist.

3.3.3.3 Minkowski Error Function

The Minkowski EF is characterized by a parameter that appears as the index in the definition. The additional feature of the function makes it wider than it actually is, in that the other standard functions studied thus far (like absolute error, fourth power function, quadratic error) can be obtained as particular cases of this function (by accordingly setting the index). For even indices the Minkowski function behaves like the quadratic function typically (for all others in this class have similar properties like convexity toward x-axis, symmetry about the y-axis). The odd indices generate functions that need a piece-wise defined update rule for the function would be symmetric about the origin in this case. The complex Minkowski EF keeps the form of the actual real valued function.

3.3.3.4 Mean-Median Error Function

The mean-median EF behaves like absolute function for large values of the error. The function is convex through out the x-axis. For smaller error values, the function behaves like quadratic EF. The function finds best application for data that are prone to an outlier scatter that should be treated by assigning a function to nullify the ill effects due to them. Since the mean-median EF behaves like absolute function that allots less cost that the conventional quadratic function would have, the solution obtained by this function would be better than the one obtained by employing quadratic function. The complex version of the mean-median EF carries the above-described features forward to the complex plane. The extended function hence, behaves like the quadratic EF for smaller values of the complex error. For complex errors large, the function is remnant of the Absolute Error.

3.3.3.5 Sine-Hyperbolic Error Function

The sinh EF is symmetric about the origin. The computation of slope should be directed toward the origin in the first and the third quadrants separately, for which the update rule must be defined in these quadrants accordingly. The function is smooth and maintains convexity through out the x-axis. The slope computed from this EF varies according to cosh(x). The complex Sinh EF extends the hyperbolic sine function to the complex domain. A complex conjugation was employed in the argument to the function that makes the extended complex function an even function. This is rotationally symmetric about the z-axis. The surface maintains convexity with respect to xy-plane. The steepness of the slope increases as the index rises.

3.3.3.6 Cauchy Error Function

The Cauchy EF has one minimum point at the origin. The function is symmetric about the y-axis. The training steers the weights so as to reach the minimum of the function. The function is defined through out the real line, it is continuous everywhere and differentiable all through. The function changes convexity as x increases. As a ramification of this fact, the update from the slope function based on this EF at larger values of x would be smaller in comparison with that of the quadratic EF. The complex Cauchy EF was defined to perform the Cauchy function for the complex variables. The surface plot reveals a unique point of global minimum. The surface is differentiable through out the real plane. The surface is also characterized by changing convexity as the radius vector increases. It is rotationally symmetric about the z-axis.

3.3.3.7 Huber Error Function

The definition of Huber EF has both features of the quadratic error and the absolute error. The function enables one to optimally choose EFs. If the data were prone to outliers and if their scatter is biased to one side, an obvious choice would be to suppress the influence of these spurious points by assigning an absolute EF to the side and set quadratic function to operate on the other side. It was found that in statistical analysis such choice indeed bettered the results as a judicious assignment was proven to be effective. The complex Huber function was defined to generalize the real Huber function to the complex variables retaining the form the function. The principle of operation of the function remains the same as the choice of the parameter c assigns the domains of operation of the EF. The quadratic and the absolute functions exist in the definition and the assignment will accordingly suppress the ill-effects of the outliers and other spurious data by evaluating the assigned function accordingly. It must be noted that the extended function is a paraboloid of revolution for the part of the definition that was quadratic and for the part that was absolute function, the extended version is a cone.

3.3.3.8 Welsch Error Function

The Welsch EF is designed to suppress large errors and give a quadratic function like performance in the vicinity of the origin. The function is convex with respect to x-axis near the origin and has the asymptote $y = c^2/2$. The complex version of the function retains the form but operates with complex errors. Convexity prevails near the origin, the plane $z = c^2/2$ is an asymptote to the surface. On one hand, it suppresses large errors while on the other hand it gives an quadratic function like performance for small errors.

These error functions considered in this chapter are collected from various sources and refined into the neural network prospective. Taking this as the point of start, the basic question needs attention from the viewpoint of the deriving learning rule for BP and the ***C***BP. The interested readers may easily obtain the learning rule for corresponding EF based backpropagation algorithm by associating the definition of EF in Eq. 3.27. The derivatives of these functions may be easily computed to implement the basic update rule for training the ***R***VNN and ***C***VNN. As a ramification of the fact, the additional factor that enters the weight update rule in the BP and the ***C***BP has the same form. How the BP and the ***C***BP would perform when the EF is varied, validate the performances by applying them to some well-known benchmarks.

3.4 Learning in Complex Domain

The learning rules correlate the input-output mapping of the nodes by iteratively adjusting the interconnection weights. It can be seen that the learning algorithm in complex domain is steadily gaining prominence yet is in an embryonic stage. The avenues for the BPA further open up as the survey indicates but once established the ***C***BP can compete with the BPA in problems where both could be applied. Needless to state that the ***C***BP would be preferred over BPA in applications that demand the real and imaginary parts of complex numbers and functions be retained and no modeling involving a tailoring with these quantities may be allowed. Such applications require that the physical significance of the complex numbers be kept intact. Typically, signal processing is one area where such requirements exist. The standard real back-propagation (***RBP***) learning is a widely used straight forward algorithm, but it has limitations such as slow convergence, may even get trapped in local minima and low degree of accuracy in many cases. However, the ***C***BP algorithm improves these issues considerably.

Various gradient-based learning algorithms in complex domain had been developed [7, 11, 17, 19, 21] in last few years. The theoretical aspects of these algorithms have diverse viewpoints depending upon the complex-valued nonlinear activation functions used [24]. The characteristics of complex nonlinearities and associated learning algorithms are related to the distinguishing features of complex-valued nonlinear adaptive filtering [25]. It is worth mentioning that complex back-propagation (***C***BP) algorithm reduces the probability for standstill in learning and improves the learning speed considerably [1, 7]. The performance of ***C***BP has been found far superior for complex-valued [1, 22, 26] as well as for real-valued [3, 8, 24] problems.

It is imperative to consider an automated and robust learning method with a good performance and free from limitations of BP. In this chapter, a modified real-resilient propagation (***RRPROP***) learning algorithm proposed in [27] has been extended in complex domain. Moreover, an improved complex-resilient propagation (***C****RPROP*) algorithm with error dependent weight backtracking step has also been presented for efficient learning. These learning algorithms have been defined with bounded but nonanalytic complex activation function.

3.4.1 Complex Back-Propagation (CBP) Learning

The ***C***BP algorithm for CVNN have been proposed by several researchers in recent years [1]. The ***C***BP is a complex domain version of ***RBP***. The aim is to approximate a function which will map the inputs to the outputs by using a finite set of learning patterns (z, y).

$$y = f(z, w)$$

where $w \in C$ corresponds to all weights and thresholds in neural networks, $z \in \boldsymbol{C}$ corresponds to all complex valued training input patterns, and $y \in \boldsymbol{C}$ corresponds to all complex valued training output patterns. There are two broad classes in the C-BP proposed by several researchers. One is based on the activation function which maps complex number to complex number through a complex function. In the other approach, the complex variable is split into two parts, real and imaginary, and then the activation function is separately applied to these parts to obtain the real and imaginary parts of the complex output.

In order to directly process complex signals by an artificial neural networks, various gradient-based learning algorithms in complex domain had been developed in [7, 11, 19, 21]. The theoretical aspects of these algorithms have diverse viewpoints depending upon the complex-valued activation functions used, as explained in previous section. The derivation of learning rules in this chapter are based on split-type activation function, as given in Eq. 3.3. The ***C***BP with this activation function compromises the analytic property of the activation function for boundedness. With split activation function, the update rules are linear combination of derivatives of real and imaginary components of an activation function. Both real and imaginary parts of the weights are modified as the function of real and imaginary parts of signals [9]. This structure reduces the probability of standstill [7, 9] in $\boldsymbol{C}BP$ as compared to $\boldsymbol{R}BP$ and enhances average learning speed. The unit of learning is complex valued signal and learning in complex domain neural network is adjusting 2D motion [7].

The gradient descent-based error back-propagation is a very popular learning procedure for feed-forward neural networks. This conventional back-propagation learning algorithm in real domain (***RBP***) have been extended to (***CBP***). Let $e_n = D_n - Y_n$ be the difference between actual (Y_n) and desired (D_n) output of nth neuron in output layer. The real-valued cost function (MSE) can be given as:

$$E = \frac{1}{2N} \sum_{n=1}^{N} |e_n|^2 = \frac{1}{2N} \sum_{n=1}^{N} \left[(\Re(e_n))^2 + (\Im(e_n))^2 \right] \tag{3.25}$$

Cost function E is a scalar quantity which $\mathbb{C}$BP algorithm minimizes by modifying the weights and biases of the network. As this function is a real-valued nonanalytic function, the partial derivative of E with respect to the real and imaginary part of the weights and biases are found separately. The weights and biases are recursively altered by applying gradient descent on the energy function E, given by,

$$w^{\text{new}} = w^{\text{old}} - \eta \ \nabla_w \ E, \tag{3.26}$$

where, the gradient $\nabla_w E$ is derived with respect to both real and imaginary parts of complex weights. The weight update is proportional to the negative of gradient, hence

$$\begin{aligned} \Delta w = -\eta \ \nabla_w \ E &= -\eta \ \nabla_{\Re(w)} \ E - j \ \eta \ \nabla_{\Im(w)} \ E \\ &= -\eta \left(\frac{\partial E}{\partial \Re(w)} + j \times \frac{\partial E}{\partial \Im(w)} \right) \end{aligned} \tag{3.27}$$

3.4.2 Complex Resilient Propagation (CRPROP) Learning

The ***RBP*** algorithm is widely used for training neural networks because of ease in implementation. However, the slow rate of convergence and getting stuck into local minima are the major limitations in the algorithm performance. Further, some modifications and variations to the basic error back-propagation procedure, like addition of momentum term, modified EF [28, 29], Delta-Bar-Delta algorithm [30] and Quick Prop [31] were suggested to overcome these problems. However, none of these modifications have accelerated the rate of convergence to a large extent. For efficient learning and faster convergence, resilient propagation in real domain (***RRPROP***) [32] was proposed. The basic principle in this algorithm is to eliminate the harmful influence of the size of partial derivatives of EF on weight update and adaptation is made dependent on the sign of the derivative. Further modification in real resilient propagation without increasing the complexity of algorithm was suggested [27] for significant improvement in the learning speed. This local gradient based adaptation technique can learn quickly with clearly lesser number of computations. Resilient back-propagation (*RPROP*) is a local adaptive learning scheme, performing supervised batch learning in multi-layer neural network. It is basically aimed at eliminating the harmful influence of the size of the partial derivative on the weight step. In RPROP, only the sign of the derivative is considered to indicate the direction of the weight update and the size of the weight change is exclusively determined by a weight-specific update value. The complex-RPROP algorithm can be derived by extending the real RPROP to the complex domain.

The weight update in back-propagation learning depends on the size of partial derivative of EF, while in resilient propagation (RPROP) it is independent of the size of partial derivative. The weight update in RPROP depends only on the temporal behavior of its sign. This result in faster convergence of the learning process [32]. The ***CRPROP*** algorithm has been derived by extending the real RPROP given in [27] to complex domain. The aim of the ***CRPROP*** algorithm is to modify the real and imaginary parts of the complex weights by an amount $\Re(\Delta(t))$ and $\Im(\Delta(t))$ (update values), in such a way so as to decrease the overall error. These update values solely determine the size of weight update $\Delta\Re(w(t))$ and $\Delta\Im(w(t))$ and the sign of partial derivative $(\partial E/\partial\Re(w))$ and $(\partial E/\partial\Im(w))$ determine the direction of each weight update. $(\partial E/\partial\Re(w))$ and $(\partial E/\partial\Im(w))$ are the gradient summation over all patterns of the pattern set. The update value is initialized to Δ_0 and then it is modifying according to gradient direction, as given in algorithm. Some other parameters are also set at the beginning of resilient propagation algorithm. They are increase factor (μ^+), decrease factor (μ^-), minimum step size (Δ_{min}) and maximum step size (Δ_{max}). The weight updates are computed as follows:

$$\Re(\Delta w(t)) = -\text{sign}\left(\frac{\partial E(t)}{\partial\Re(w)}\right)\Re(\Delta(t)) \tag{3.28}$$

$$\Im(\Delta w(t)) = -\text{sign}\left(\frac{\partial E(t)}{\partial\Im(w)}\right)\Im(\Delta(t)) \tag{3.29}$$

In other words, every time the partial derivative of the corresponding weight w changes its sign, which indicates that the last update was too big and the update value is decreased by the factor μ^-. If the derivative retains its sign, the update value is slightly increased by factor μ^+, in order to accelerate the convergence. So there is effectively growing and shrinking of the update value according to the sign of the gradient. The factors μ^- and μ^+ are set to the 0.5 and 1.2, respectively.

3.4.3 Improved Complex Resilient Propagation (C-iRPROP)

The change of sign of the partial derivative in successive steps is considered as a jump over minima, which results in reversal of previous weight update [32]. This decision does not take into account whether the weight update has caused an increase or decrease of error. This backtracking step does not seem proper especially when the overall error has decreased. Therefore, here we present the ***CRPROP*** *algorithm with error dependent weight reversal step* and the resulting algorithm has shown excellent performance. Assuming that the network is close to the (local) minimum, the each weight update not leading to a change of sign of the corresponding partial derivative leads the algorithm closer to the optimum value. Hence, previous weight update is reverted only when it has caused the change of sign in corresponding partial derivative in case of an overall error increase:

$$\text{if}\left(\frac{\partial E(t-1)}{\partial \Re(w)} \times \frac{\partial E(t)}{\partial \Re(w)} < 0 \text{ and } E(t) > E(t-1)\right)$$
$$\text{then } \Re(w(t+1)) = \Re(w(t)) - \Re(\Delta w(t-1)) \tag{3.30}$$
$$\text{if}\left(\frac{\partial E(t-1)}{\partial \Im(w)} \times \frac{\partial E(t)}{\partial \Im(w)} < 0 \text{ and } E(t) > E(t-1)\right)$$
$$\text{then } \Im(w(t+1)) = \Im(w(t)) - \Im(\Delta w(t-1)) \tag{3.31}$$

The initialization parameters are same as in case of conventional ***C***RPROP and the change is only in the case where partial derivative changes its size. The resulting algorithm shows better performance as compare to ***C***RPROP. Let w_{lm} be the weight from lth neuron in a layer to mth neuron in next layer in a neural network. Let t = 1 and $0 < \mu^- < \mu^+ < 1.2$. The pseudo code for this ***C***-*i*RPROP algorithm is given below:

$$\forall l, m : \Re(\Delta_{lm}(t)) = \Im(\Delta_{lm}(t)) = \Delta_0 \quad \text{and} \quad \frac{\partial E(t-1)}{\partial \Re(w_{lm})} = \frac{\partial E(t-1)}{\partial \Im(w_{lm})} = 0$$
$$\Re(\Delta_{\max}) = \Im(\Delta_{\max}) = \Delta_{\max} \quad \text{and} \quad \Re(\Delta_{\min}) = \Im(\Delta_{\min}) = \Delta_{\min}$$

Repeat {

calculate $\frac{\partial E(t)}{\partial \Re(w_{lm})}$ and $\frac{\partial E(t)}{\partial \Im(w_{lm})}$ for all weights and biases

For real part of weight:

$$\text{if}\left(\frac{\partial E(t-1)}{\partial \Re(w_{lm})} \times \frac{\partial E(t)}{\partial \Re(w_{lm})} > 0\right) \text{ then}$$
$$\{\Re(\Delta_{lm}(t)) = \min(\Re(\Delta_{lm}(t-1)) \times \mu^+, \Re(\Delta_{\max}))$$
$$\Re(\Delta w_{lm}(t)) = -\text{sign}\left(\frac{\partial E(t)}{\partial \Re(w_{lm})}\right) \Re(\Delta_{ij}(t))$$
$$\Re(w_{lm}(t+1)) = \Re(w_{lm}(t)) + \Re(\Delta w_{lm}(t))\}$$

$$\text{if}\left(\frac{\partial E(t-1)}{\partial \Re(w_{lm})} \times \frac{\partial E(t)}{\partial \Re(w_{lm})} < 0\right) \text{ then}$$
$$\{\Re(\Delta_{lm}(t)) = \max(\Re(\Delta_{lm}(t-1)) \times \mu^-, \Re(\Delta_{\min}))$$
$$\text{if } (E(t) > E(t-1)) \text{ then}$$
$$\Re(w_{lm}(t+1)) = \Re(w_{lm}(t)) - \Re(\Delta w_{lm}(t-1)) \quad \text{and} \quad \frac{\partial E(t)}{\partial \Re(w_{lm})} = 0\}$$

$$\text{if}\left(\frac{\partial E(t-1)}{\partial \Re(w_{lm})} \times \frac{\partial E(t)}{\partial \Re(w_{lm})} = 0\right) \quad \text{then}$$
$$\{\Re(\Delta w_{lm}(t)) = -sign\left(\frac{\partial E(t)}{\partial \Re(w_{lm})}\right) \Re(\Delta_{ij}(t))$$
$$\Re(w_{lm}(t+1)) = \Re(w_{lm}(t)) + \Re(\Delta w_{lm}(t))\}$$

For Imaginary part of weight:

$$\text{if}\left(\frac{\partial E(t-1)}{\partial \Im(w_{lm})} \times \frac{\partial E(t)}{\partial \Im(w_{lm})} > 0\right) \text{ then}$$
$$\{\Im(\Delta_{lm}(t)) = \min(\Im(\Delta_{lm}(t-1)) \times \mu^{+}, \Im(\Delta_{\max}))$$
$$\Im(\Delta w_{lm}(t)) = -\text{sign}\left(\frac{\partial E(t)}{\partial \Im(w_{lm})}\right) \Im(\Delta_{ij}(t))$$
$$\Im(w_{lm}(t+1)) = \Im(w_{lm}(t)) + \Im(\Delta w_{lm}(t))\}\}$$

$$\text{if}\left(\frac{\partial E(t-1)}{\partial \Im(w_{lm})} \times \frac{\partial E(t)}{\partial \Im(w_{lm})} < 0\right) \text{ then}$$
$$\{\Im(\Delta_{lm}(t)) = \max(\Im(\Delta_{lm}(t-1)) \times \mu^{-}, \Im(\Delta_{\min}))$$
$$\text{if } (E(t) > E(t-1)) \text{ then}$$
$$\Im(w_{lm}(t+1)) = \Im(w_{lm}(t)) - \Im(\Delta w_{lm}(t-1)) \quad \text{and} \quad \frac{\partial E(t)}{\partial \Im(w_{lm})} = 0\}$$

$$\text{if}\left(\frac{\partial E(t-1)}{\partial \Im(w_{lm})} \times \frac{\partial E(t)}{\partial \Im(w_{lm})} = 0\right) \quad \text{then} \quad \{\Im(\Delta w_{lm}(t)) = -\text{sign}\left(\frac{\partial E(t)}{\partial \Im(w_{lm})}\right) \Im(\Delta_{ij}(t))$$
$$\Im(w_{lm}(t+1)) = \Im(w_{lm}(t)) + \Im(\Delta w_{lm}(t))\}\}$$

$t = t + 1\,\}$ **Until(converged)**

The update values and the weights are changed every time when a new training set is presented. All update values (Δ_{lm}) are initialized to Δ_0. The initial update value, Δ_0, is selected in a reasonably proportion to the size of initial weights. In order to prevent the weights from becoming too small and too large, the range of update value has been restricted to a minimum limit ($\Delta_{\min}$) and maximum limit ($\Delta_{\max}$). In experiments, it had been throughly seen that by setting these update value quiet small, one could obtain a smooth learning process. The choice of decrement factor $\mu^{-} = 0.5$ and increment factor $\mu^{+} = 1.2$ generally yields good results. It was also observed that small variation in these values did neither improve nor deteriorate learning process.

3.5 Concluding Remarks

The approach in the chapter went along the following lines: The two important parameters that exist in the neural network learning viz Activation Function and Error Function were thoroughly discussed with associated properties. The chapter is further extended to an improved and fast leaning algorithm which counters the ill effects of popular back-propagation learning algorithm in complex domain. Of the various parameters set to run the CVNN, the $\mathbf{C}$AF is most critical. In the complex domain, analyticity of the functions is supposed to be verified before it can be used as an activation function to the CVNN. This aspect is unlike real-valued ANN where no such constraint existed and AF chosen to be a smooth (continuously differentiable) and bounded function. On the other hand, the complex plane imposes its own constraint in the form of Liouville Theorem, which states that if a complex-valued function is both analytic and bounded, it must be a constant function. It is useless to consider a constant function for neuron's activation to accomplish learning process therefore many researchers have favored for the boundedness of $\mathbf{C}$AF . The Liouville Theorem constraint is studied here and the different $\mathbf{C}$AFs proposed are surveyed. A study into the boundedness behavior of the so-called "split" type function for activation of neuron is revealed in cost of its analytic behavior. The broad picture of $\mathbf{C}$AFs available from literature have been collected and analyzed. The $\mathbf{C}$AFs, Haykin Activation and Split Activation were comparatively explained. The singular points of the Haykin fully complex-valued activation function, which were found to be responsible for disrupting the downstream convergence whenever at least one of the net inputs to neurons fell in their vicinity in the course of training. In 2003, Adeli [11] thoroughly explored fully complex-valued function for CVNN and stated that the universality of the networks can be shown by dealing with these singularities, such that singularities are removed or avoided by restricting the regions. The split-type function for activation found to be a better choice for construction of CVNN in all respects.

This chapter presents prominent EFs as a basis for making these choices and designing a learning scheme. This approach will clearly offer more tooling while designing neural networks as firstly, it departs from the present day technique of using the quadratic EF and invoking the procedure of weight update using a gradient descent, which in practice does not manage to go below a certain value of the error (although theory assures that an neural network exists for an arbitrary error that the investigator may desire). The study reveals that the EF can indeed be treated as a parameter while employing learning in RVNN or CVNN for training of presented data. The deployment of other EFs that may perform well and can replace the quadratic EF is contingent upon the applications or requirements. It must be noted that in the description of EFs and the function's form has been retained while extending to the complex domain. This was done to make sure that the error computed kept the same formula and also makes sure that the surface plot of the function is close to the plane plot of the same, even while operating in the complex domain.

The main drawback of BP is the large computational complexity due to extensive training. One way to decrease the computational load in multilayer network is to fasten the training process. The resilient propagation algorithm in real and complex domain are well tried by neural network community. The modification in existing complex resilient propagation algorithm possesses reasonable logic which improves its learning speed without increasing the complexity of algorithm. The wide spectrum of benchmark and reallife examples presented in successive chapters confirm the motivation of improved ***C***RPROP. For a fair comparison, the computational costs for optimization are measured in number of learning cycles [27]. The improved complex resilient propagation (***C-i***RPROP) can provide much faster convergence in comparison to ***C***BP and thus also smaller computational complexity.

References

1. Hirose, A.: Complex-Valued Neural Networks. Springer, New York (2006)
2. Nitta, T.: Orthogonality of decision boundaries in complex-valued neural networks. Neural Comput. **16**(1), 73–97 (2004)
3. Aizenberg, I., Moraga, C.: Multilayer feedforward neural network based on multi-valued neurons (MLMVN) and a back-propagation learning algorithm. Soft Comput. **11**(2), 169–183 (2007)
4. Amin, M.F., Murase, K.: Single-layered complex-valued neural network for real-valued classification problems. Neurocomputing **72**(4–6), 945–955 (2009)
5. Savitha, R., Suresh, S., Sundararajan, N., Kim, H.J.: Fast learning fully complex-valued classifiers for real-valued classification problems. In: Liu, D., et al. (eds.) ISNN 2011, Part I, Lecture Notes in Computer Science (LNCS), vol. 6675, pp. 602–609 (2011)
6. Savitha, R., Suresh, S., Sundararajan, N., Kim, H.J.: A fully complex-valued radial basis function classifier for real-valued classification. Neurocomputing **78**(1), 104–110 (2012)
7. Nitta, T.: An extension of the back-propagation algorithm to complex numbers. Neural Netw. **10**(8), 1391–1415 (1997)
8. Tripathi, B.K., Kalra, P.K.: Complex generalized-mean neuron model and its applications. Appl. Soft Comput. (Elsevier Science) **11**(1), 768–777 (2011)
9. Nitta, T.: An analysis of the fundamental structure of complex-valued neurons. Neural Process. Lett. **12**, 239–246 (2000)
10. Savitha, R., Suresh, S., Sundararajan, N., Saratchandran, P.: A new learning algorithm with logarithmic performance index for complex- valued neural networks. Neurocomputing **72**(16–18), 3771–3781 (2009)
11. Kim, T., Adali, T.: Approximation by fully complex multilayer perceptrons. Neural Comput. **15**, 1641–1666 (2003)
12. Savitha, R., Suresh, S., Sundararajan, N.: A fully complex-valued radial basis function network and its learning algorithm. Int. J. Neural Syst. **19**(4), 253–267 (2009)
13. Amin, M.F., Islam, M.M., Murase, K.: Ensemble of single-layered complex-valued neural networks for classification tasks. Neurocomputing **72**(10–12), 2227–2234 (2009)
14. Li, M.-B., Huang, G.-B., Saratchandran, P., Sundararajan, N.: Fully complex extreme learning machine. Neurocomputing **68**, 306–314 (2005)
15. Brown, J.W., Churchill, R.V.: Complex Variables and Applications, VIIth edn. Mc Graw Hill, New York (2003)
16. Saff, E.B., Snider, A.D.: Fundamentals of Complex Analysis with Applications to Engineering and Science. Prentice Hall, Englewood Cliffs (2003)
17. Piazza, F., Benvenuto, N.: On the complex backpropagation algorithm. IEEE Trans. Sig. Proc. **40**(4), 967–969 (1992)
18. Tripathi, B.K., Kalra, P.K.: The novel aggregation function based neuron models in complex domain. Soft Comput. (Springer) **14**(10), 1069–1081 (2010)

19. Georgiou, G.M., Koutsougeras, C.: Complex domain backpropagation. IEEE Trans. Circuits Systems-II: Analog Digital Signal Proc. **39**(5), 330–334 (1992)
20. Widrow, B., McCool, J., Ball, M.: The complex LMS algorithm. Proc. IEEE **63**(4), 719–720 (1975)
21. Leung, H., Haykin, S.: The complex backpropagation algorithm. IEEE Trans. Signal Proc. **39**(9) (1991)
22. Tripathi, B.K., Kalra, P.K.: Functional mapping with complex higher order compensatory neuron model. In: World Congress on Computational Intelligence (WCCI-2010). ISSN: 1098–7576. IEEE Xplore, Barcelona, Spain, 18–23 July 2010
23. Kim, M.S., Guest, C.C.: Modification of backpropagation networks for complex-valued signal processing in frequency domain. In: Proceedings of IJCNN, San Diego (1990)
24. Tripathi, B.K., Kalra, P.K.: On efficient learning machine with root power mean neuron in complex domain. IEEE Trans. Neural Netw. **22**(05), 727–738, ISSN: 1045–9227 (2011)
25. Mandic, D., Goh, V.S.L.: Complex valued nonlinear adaptive filters: noncircularity, widely linear and neural models. Wiley, Hoboken (2009)
26. Shin, Y., Jin, K.-S., Yoon, B.-M.: A complex pi-sigma network and its application to equalization of nonlinear satellite channels. In: IEEE International Conference on Neural Networks (1997)
27. Igel, C., Husken, M.: Empirical evaluation of the improved Rprop learning algorithms. Neurocomputing **50**, 105–123 (2003)
28. Van Ooyen, A., Nienhuis, B.: Improving the convergence of the back-propagation algorithm. Neural Netw. **5**(3), 465–472 (1992)
29. Chen, L., Zhang, L., Zhang, H., Abdel-Mottaleb, M.: 3D Shape constraint for facial feature localization using probabilistic-like output. In: Proceeding of 6th IEEE International Conference on Automatic Face and Gesture Recognition (2004)
30. Jacobs, R.A.: Increased rates of convergence through learning rate adaptation. Neural Netw. **1**(4), 295–307 (1988)
31. Fahlman, S.E.: An empirical study of learning speed in back-propagation networks. In: Technical report, CMU-CS-88-162 (1988)
32. Riedmiller, M., Braun, H.: A direct adaptive method for faster back-propagation learning: the RPROP algorithm. In: Proceeding of IEEE ICNN'93, San Francisco, pp. 586–591 (1993)

Chapter 4
Higher-Order Computational Model for Novel Neurons

Abstract Artificial neural network (ANN) has attracted a tremendous amount of interest for the solution of many complicated engineering and real-life problems. A small complexity, quick convergence, and robust performance are vital for its extensive applications. These features are pertinent upon the architecture of the basic working unit or neuron model, used in neural network. The computational capability of a neuron governs the architectural complexity of its neural network, which in turn defines the number of nodes and connections. Therefore, it is imperative to look for some neuron models, which yield ANN having small complexity in terms of network topology, number of learning parameters (connection weights) and at the same time they should possess fast learning, and superior functional capabilities. The conventional artificial neurons compute its internal state as the sum of contributions (aggregation) from impinging signals. For a neuron to respond strongly toward correlation among inputs, one must include higher-order relation among a set of inputs in their aggregation. A wide survey into design of artificial neurons brings out the fact that a higher-order neuron may generate an ANN which can have better classification and functional mapping capabilities with comparatively less number of neurons. Adequate functionality of ANN in a complex domain has also been observed in recent researches. This chapter presents higher-order computational models for novel neurons with well-defined learning procedures. Their implementation in a complex domain will provide a powerful scheme for learning input/output mapping in complex as well as in real domain along with better accuracy in wide spectrum of applications. The real domain implementation may be realized as its special case. The purpose of investigation in this chapter is to present the suitability and sustainability of higher-order neurons for readers, which can serve as a basis of the formulation for powerful ANN.

4.1 Biological Neuron

The human nervous system is a tremendously complicated structure consisting of about 10^{11} neuron units. The basic building block of biological information processing system, the neuron, consists of three basic components viz dendrites, soma, and

B.K. Tripathi, *High Dimensional Neurocomputing*,
Studies in Computational Intelligence 571, DOI 10.1007/978-81-322-2074-9_4

axon; and are arranged in functional constellation or assemblies according to the synaptic contacts they make with one another. The dendrites, highly branched construction, are receptive surfaces for input signals and conduct them to neuron cell. Aggregation of these signals takes place in cell body (soma) in unique fashion whose characteristic derive the computing capability of the neuron. Axon, that begins with axon hillock, which generates the cell action potential and converts it into a train of impulses through itself for transmission. In order to convey the action potential, the dendrites of one neuron are connected with the axon of the other neurons via synaptic connections or synapses. The synaptic transmission involves complicated electrical and chemical processes in the system.

The spatial integration (aggregation) of the synapses on the dendritic tree is reflected in the computations performed at local branches [1–3]. The linear interaction (summation) of synaptic inputs is popularly believed for long time to model the computation abilities of neurons. Linearity is believed to be sufficient in capturing the passive properties of dendritic membrane where synaptic inputs are currents. However, many researches divulge the nonlinear interaction of synaptic inputs in cell body for information processing. The multiplication of inputs is also found to exist in real nervous systems by animals and in other biological evidences. The synaptic inputs can interact nonlinearly when synapses are co-localized on patches of dendritic membrane with specific properties. A Multiplication, the most basic of all nonlinearity in neurons, often occurs in dendritic trees with voltage-dependent membrane conductance [2]. But it is not very clear that how real neurons do the multiplication, where as it is clear that multiplication of the inputs of neuron can increase the computational power and storage capacities of neural networks. Therefore, an artificial neuron model should then be capable of including this inherent nonlinearity in the mode of aggregation.

4.2 Artificial Neuron

An artificial neuron is a mathematical representation of the biological neuron which tries to approximate its functional capabilities. A neuron model is characterized by its formalism and its precise mathematical definition. Historically, McCulloch and Pitts model [4] is supposed to be first formal mathematical model of an artificial neuron based on the highly simplified consideration of biological neuron. This conventional neuron possesses the weighted summation (as aggregation) of impinging signals and neglects all possible nonlinear capabilities of the single neuron.

An important issue in artificial neuron model is the description of single neuron computation and interaction among the input signals. A neuron model can be described as a combination of aggregation and activation functions. The net potential of a neuron is characterized by an aggregation function, which models the integration of impinging information. The activation function limits the amplitude of the neuron output to some finite range. In order to qualitatively describe the functional description of any neuron model, one needs to know how the input signals are aggregated and processed to obtain the output.

ANN, a simplified model of the biological neural network, is a massively parallel distributed processing system which is made up of highly interconnected neural computing units (neurons) having ability to learn and generalize. Generalization refers to the neural network producing appropriate outputs for inputs not encountered during training. The learning and generalization capabilities of neural networks make them possible to solve the complex problems which are currently intractable. The important process in artificial neuron is forming a unit net potential from impinging signals. There are various ways to aggregate the input values to get the unit potential value like additive, subtractive, multiplicative, polynomial, rational, etc. The summation of inputs in neuron played a vital role in construction of an artificial neuron. Such neuron model when used to solve the real-life problem may require a large number of neurons in the network which means that the complexity of the network and its computational burden will be extensively increased. This problem more worsens when dealing with high- dimensional applications. This problem can be overcome by the consideration of various factors, such as architecture, node's functionality, learning rule, and training sets. Their appropriate choice is necessary for an efficient design of the ANN. But, there are three leading subjects to trim down the complexity and computational burden of network along with efficient learning and superior results.

- *Reducing the number of neurons in ANN by introducing higher-order neurons.* The higher-order neuron may produce better learning and generalization performance with reasonably less number of connected nodes.
- *Fasten the training process by selecting the efficient learning algorithm.* The learning speed of ANN depends upon the nature of the learning algorithm. Apart from straightforward, simple, and conventional backpropagation algorithm there are many learning algorithm which provide much faster convergence.
- *Implementation of neural networks in higher dimensions.* The learning and architectural complexity of ANN also depends upon the complexity of the application concerned. When dealing with high-dimensional data, it is better to consider neurons which accepting high-dimensional data as single cluster. This drastically reduces the number of connected nodes as compared in the conventional ANN.

In the last decade, many researches in the field of ANN have been directed toward the evolution of superior architecture for neural system for efficient learning and better analysis of high-dimensional data. This chapter is mainly concerned with mathematical modeling of higher-order neuron and their implementation in complex domain, which is an important aspect in studying the fundamental principal of information processing. Three example neuron models are presented here to demonstrate the motivation of introduction of higher-order neuron modeling.

4.2.1 Higher-Order Aggregation Function

The human nervous system consists of neuron cells possessing complex structure with highly complex physiological properties and operation. Any system level investigation of the nervous system would demand an explanation on the information processing in the constituent neurons. This requires a description of how the input signals to the neuron interact and jointly affect the processing. Neuron computation involves complex processing of a large number of signals before resulting in the output signals. In numerous studies [3, 5], it was found that the computational power of a neuron cell lies in the way of aggregation of synaptic signals in cell body. Attempts to give a mathematical representation to the aggregation process of synapses continue to be a fascinating field of work for researchers in the neurocomputing community. A neuron is a basic processing unit of an ANN which is characterized by a well-defined aggregation function of weighted inputs.

An artificial neuron is the simplified model of a biological neuron which can approximate its functional capabilities. But, for the time being, it is far from clear how much of this simplicity is justified, since at present we have only a poor understanding of neuronal functions in biological networks. A neuron cell in nervous system has very complex structure and extremely complex physiological properties and operations. An artificial neuron in the neural network is usually considered as computational model of a biological neuron, i.e. real nerve cell, and the connection weights between nodes resemble to synapses between the neurons. The computational power of a neuron is a reflection of spatial aggregation of input signals in the cell body. This chapter is focused on the design and assessment of nonlinear aggregation functions for artificial neurons.

The very generic organism of neuron computation is to use the summation aggregation function. C Koch and T Poggio (1992) explain the relevance of using multiplication as a computationally powerful and biologically realistic possibility in computation [1]. In neurophysiology, the possibility that dendritic computations could include local multiplicative nonlinearities is widely accepted. Mel and Koch [6] argued that sigma-pi units underlie the learning of nonlinear associative maps in cerebral cortex. This leads us to develop a very flexible aggregation function that is biologically plausible and more powerful than conventional one. This chapter presents three new higher-order aggregation functions and exploit the ways in generating the comprehensive neural units. Figure 4.1 portrays a flexible artificial neuron with aggregation function Ω.

4.2.2 Why Higher-Order Neurons

A brief survey into wide applications of neural networks points to the fact that, when the desired mapping is complicated and input dimension is high, it is tough to foresee how long the learning process of the neural network of conventional neuron will take and weather the learning will converge to an satisfactory result. The

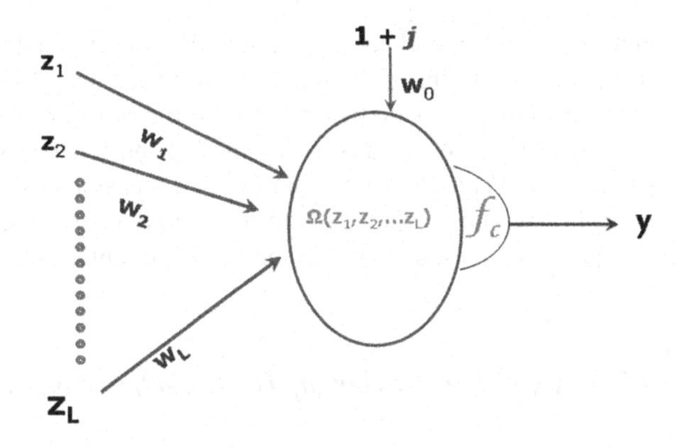

Fig. 4.1 An artificial neuron model with aggregation function Ω

use of neural networks with conventional neurons is naturally accompanied by the trade-off between issues, such as the complexity of network, overfitting, generalization capability, local minima problems, and stability of the weight update system. In recent past, much attention has been given to convergence and stability evaluation of neural network weights by energy function-based optimization methods, to training techniques to avoid weights from getting stuck into local minima and to avoid overfitting. Mostly, those techniques are sophisticated; and require substantial and time-consuming (hence costly) effort from users who are not true experts in the neural network field.

In order to overcome with these issues, one reasonable approach is to be concerned with neuron modeling which relates the structure of a neuron with its aggregation function. Neurons are functional units and can be considered as generators of function spaces over impinging signals. The major issue in artificial neuron model is the description of signal aggregation which reveals the power of single neuron computation. It was observed in many researches [5, 7, 8] that the functional capability of a neuron lies in the spatial integration of synaptic inputs. Probably not much attempts have been made for the evolution of more powerful neuron models to enhance the overall performance and computational capabilities of ANN. A neuron designed with higher-order correlation over impinging signals leads to a superior mapping and computing capabilities [9]. Therefore, multiplication units [7] have became a natural choice in modeling a computationally powerful and biologically plausible extension to conventional neurons. Further, the most well-known units that comprise of multiplicative synapses are perhaps higher-order neurons; which have been developed to enhance the nonlinear expressional ability of the multilayer neural networks. Motivated by the higher-order characteristics of the neuron and the classic Stone-Weierstrass's theorem [10], a class of neuron models known as pi-sigma [11], sigma-pi [12], polynomial neural networks [13], and higher-order neurons [9] have been introduced and successfully used. These models have been proved to be more efficient as both single units as well as in networks. These neuron models forms the higher-order polynomials on the basis of the number of inputs in space, improves the learning capability in terms of speed and performance with lesser number of neurons in the network as compared to the conventional neural networks. Interested readers may find that polynomial feedforward neural networks are appealing due to the trustworthy theoretical results for their universal approximation capabilities according to the Weierstrass theorem [10] and for their generalization power measured by the Vapnik-Chervonenkis (VC) dimension [14]. However, they suffer from typical curse of dimensionality due to a combinatorial explosion of terms as number of inputs increases, demanding sparseness in representation.

4.2.3 A Critical Review of Higher-Order Neuron Models

A network of artificial neuron is an information processing system. It provides a practical method for learning and generalization of functions, which are often difficult for a human observer to predict or understand. Conventional neurons, based on summation aggregation function, were thoroughly used to solve these problems. However, networks based on these neurons take a large number of neurons which increases the complexity of the topology, memory requirement, slow convergence, stability of the weight update system, and training time in experiments. These issues become more significant when problems considered are in high dimensions. As stated earlier, an influential framework for dealing the ANN may be characterized at three levels. These include: computational model of neurons, learning algorithms, and domain of implementation. The architecture of a neuron decides the computational power of neuron and the number of neurons in a network in turn complexity of neural network.

The conventional neurons[1] undergo from these issue because they possess a linear correlation among input components. In numerous studies it has been observed that neuron encompassing higher-order correlation among input components outperform over conventional neuron. Multiplication, being the most basic of all nonlinearities, has become a natural choice of models trying to include higher-order correlation in aggregation function. Therefore, product units [7] appeared as computationally powerful and biologically plausible extension to conventional neurons. Over the years, a substantial body of investigations has grown which yielded the wide directions in structure of higher-order neurons, hence wide range of aggregation function comprising nonlinear correlation among input signals. But, the fundamental class of higher-order neural units acquires polynomial weighting aggregation of neural inputs. The variety of neurobiological evidences [1, 2] has also grown to support the presence of nonlinear integration of synaptic inputs in the neuron cells. In broad sense the higher-order neural networks represent the same style of computation in artificial neural networks where neurons involve polynomials, or the neurons are polynomials themselves, or synaptic connections between neurons involve higher-order terms, hence higher-order polynomials. Working on this direction, an exten-

[1] In this book, a neuron with only summation aggregation function is referred as ‘conventional’ neuron and network of these neurons as ‘MLP’.

sive attempt has been by Shin and Ghosh (1991) [11], Chen and Billings (1992) [15], Chen and Manry (1993) [13], Schmidt and Davis (1993) [16], Kosmatopoulos et al. (1995) [17], Heywood and Noakes (1996) [12], Liu et al. (1998) [18], Elder and Brown (2000) [19], Bukovsky et al. (2003) [20], Hou et al. (2007) [21] and Bukovsky et al. (2009) [8] to develop the foundation of nonconventional neural units.

Motivated by their efforts, a class of structures known as pi-sigma and sigma-pi [11, 12], functional link networks [15], polynomial neural networks [13, 18, 19], higher-order neuron [16, 17, 21], quadratic, and cubic neural units [20] have been introduced. Significant and most recent publications devoted to polynomial neural networks concepts are the works of Nikolaev and Iba [22] while most recent works that are framed within higher-order neuron can be found in [8, 23]. Although, higher-order neurons have proved to be most efficient, but they suffer from the typical curse of dimensionality due to combinatorial explosion of terms when there is increase in the number of inputs. In fact, with the increase of the input dimension, the number of parameters that is weights, increases rapidly and becomes unacceptably large for use in many situations. Consequently, typically only small orders of polynomial terms are often considered in practice. This problem worsens when higher-order neurons are implemented in complex domain. These studies inspired to design new neuron models, which capture nonlinear correlation among input components but are free from the problem of combinatorial explosion of terms as the number of inputs increases. Therefore, in this chapter three new higher-order artificial neuron models have been presented with solid theoretical foundation for their functional mapping.

4.3 Novel Higher-Order Neuron Models

Neurons are functional units and can be considered as generators of function spaces. The artificial models of neuron are characterized by their formalism and their precise mathematical definition. Starting from linear aggregation proposed by McCulloch-Pitts model [4] in 1943 and Rosenblatt perceptron [24] in 1958 to higher-order nonlinear aggregations [8, 9, 19], a variety of architectures of the neurons have been proposed in literature. It has been widely accepted that the computational power of the neuron depends on the structure of aggregation function. Higher-order neurons have demonstrated improved computational power and generalization ability. However, these units are difficult to train because of a combinatorial explosion of higher-order terms when the number of inputs to the neuron increases. This section presents three higher-order neuron models which have a simpler structure without above issue; hence need not to bother for selecting the relevant monomials or the requirement of sparseness in representation that was necessary to be imposed on the other higher-order neurons to keep learning practical. These models can also be used in the same form in networks of similar units or in combination with the traditional neuron models.

Another direction is related to the domain of implementation. It is worth mentioning here that the performance of neurons in complex domain is far superior for a two as well as single-dimension problems [25–27]. The two dimension number representation comprises of real numbers and comes with phase information embedded into it. Therefore, they are significant where it is necessary to capture phase information in the data. Complex numbers form a superset of real numbers, an algebraic structure that defines real-world phenomenon like signal amplitude and phase. These are useful in analyzing various mathematical and geometrical relationships on plane. For nearly a decade, the extension of real-valued neurons for operation on complex signals [26, 28–31] have received much attention in the field of neural networks. Complex-valued neural network provides a faster convergence with better results, reduction in learning parameters (network topology), and ability to learn two dimension motion on plane [32, 33]. The main motivation in designing of proposed neuron models is to utilize the promising capabilities of nonlinear aggregation function as well as complex numbers. In this chapter, author considers higher-order neural architectures whose boom is strengthened by the introduction of complex domain implementation and corresponding learning rules.

4.3.1 Artificial Neuron Models

In literature, we generally find neuron models that comprise of summation, radial basis, or product aggregation function, as basic unit of multilayer neural network. All these models and their networks have shown their merits as well as demerits. The MLP constructs, using summation function, a global approximation for input-output mapping, while an RBF network, using exponentially decaying localized nonlinearity, constructs local approximation to input-output mapping. This section presents two compensatory type novel aggregation functions [34] for first two artificial neurons and a root-power mean-based aggregation function [25] for third neuron. First two neuron produce the net potential as linear or nonlinear composition of basic summation and radial basis operations over a set of input signals, respectively. These neurons has a compensatory basis function, whose compensatory parameters are tuned during the task of learning to model the underlying parametric relationship between summation and radial basis operations for concerned application.

In the third neuron, the aggregation function is so general that it embraces over all the averaging (aggregation) functions between minima and maxima. The aggregation process in this neuron is based on the idea underlying the weighted root-power mean [35, 36] of all inputs in the space. The variation in generalization parameter realizes its higher-order structure. These three neuron models are inspired from the class of higher-order neurons, but these models have a simpler yet compact structure without any requirement of sparseness, which was necessary in other higher-order neurons for practical learning. In the sequel, a well-defined learning rules for a multilayer network of these neurons is developed. These neuron models can be used in their original form in a network like conventional neurons or in combination with the conventional neurons. Any of these neurons may receive a vector of input and has a complex-valued aggregator that transform the weighted input into a complex-valued net potential. The neuron also has an activation function in complex domain that gives the overall output of the single neuron system as shown in Fig. 4.1. These models may be equally used in real domain discarding the imaginary part.

4.3.2 Model-1

Neuron modeling concerns with relating to the structure of the neuron on the basis of its operation. The conventional neuron models in real and complex domain are generally based on summation and radial basis aggregation functions. The traditional MLP, uses summation function as the basis function, offers global approximation to input-output mapping, but it may be plagued by long learning time and has a tendency to get trapped at bad local minima. On the contrast, RBF network, often uses the Gaussian function as the basis function, offers local approximation to input-output mapping and often provides a faster and efficient learning. However, it will be inefficient in approximating constant-valued functions, as addressed in [37]. If a curve representing training pattern is nearly constant in an interval, it would be difficult to utilize a Gaussian function to approximate this constant-valued function. The learning convergence is quick with RBF and hence has less problems when compared to MLP, but the number of RBF neurons may become quiet large for applications containing large number input variables.

The main motivation in designing the proposed aggregation function is to take advantage of the merits of perceptron and radial basis processing. This neuron has a compensatory basis function which adaptively selects the best proportion for local and global optimization, and these proportions are later multiplied to provide a higher-order function of aggregation. Which may later fed through a state activation function to create the final output. This kind of neuron itself looks complex in the first instance but when used to solve a complicated problem it yields reasonably efficient processing as compared to conventional neurons. It is also free from the basic problem of higher-order neurons that arises in case of large number of inputs. The net potential of proposed neuron is weighted composition of summation and radial basis subfunctions. Thus input aggregation proposed for this neuron is a functional, which formulates its compensatory structure. The information processed through subfunctions is integrated in desired proportion ($\gamma : \lambda$), nonlinearly $\left(\bigotimes\right)$ in the RSP (Rbf-Summation-Product) model. The compensatory parameters γ and λ specify the contribution of radial basis and summation subfunctions to take into account the vagueness involved. With a view to achieve robust aggregation function, the parameters γ and λ are itself made adaptable in course of training.

The product or intersection is usually inserted in the formulas where one speaks about nonlinear operation, expressed here as $a \bigotimes b = 1 + a + b + ab$. The novel neuron constructed with this operation is named as RSP. Let $Z = [z_1, z_2 \ldots z_L]$ be the vector of input signals, Y be an output, and f_C be the complex-valued activation function defined in Eq. (3.3). Z^T is transpose of vector Z and $\overline{z}$ is the complex conjugate of z. $W_m^S = \left[w_{1m}^S, w_{2m}^S \ldots w_{Lm}^S\right]$ is a vector of weights from input layer

$(l = 1 \ldots L)$ to summation part of mth RSP neuron $(m = 1 \ldots M)$ and $W_m^{RB} = [w_{1m}^{RB}, w_{2m}^{RB} \ldots w_{Lm}^{RB}]$ is a vector of weights from input layer to radial basis part of mth RSP neuron. All weights, bias, and input-output signals are complex numbers. By convention w_{lm} is a weight that connects l^{th} neuron to m^{th} neuron. The net potential of this neuron is defined by following aggregation function:

$$\Omega_m(z_1, z_2 \ldots z_L) = \left[\gamma_m \times \exp\left(-\left\|Z - W_m^{RB}\right\|^2\right) \bigotimes \lambda_m \times W_m^S Z^T\right], \quad (4.1)$$

where $\|Z - W_m^{RB}\|^2 = (Z - W_m^{RB}) \times (Z - W_m^{RB})^{\aleph}$. Here superscript $\aleph$ represents the matrix complex conjugate transposition. Output of the neuron may be expressed as: $Y_m = f_C(\Omega_m(z_1, z_2 \ldots z_L))$

4.3.3 Learning Rules for Model-1

A multilayer network can be constructed by new neurons similar to network of conventional neurons. The task of learning is to tune the parameters of the operator f and to model the underlying parametric relationship between the inputs and the output through the weight parameter W. We assume that the neuron observes L input-output pairs $(z_1, y_1), \ldots (z_n, y_n)$, and generates a function space that maps the vector space Z $(z \in Z)$, into the corollary responding output space Y $(y \in Y)$. Consider a commonly used three-layer network (L-M-N). First layer has L inputs, second layer has M proposed neurons and the output layer consists of N conventional neurons. Definitely this network is used in all the applications presented in this book based on *SRP* or ***CRSP*** neuron model. Let $\eta \in [0, 1]$ be the learning rate and f' be the derivative of function f. w_0 is a bias and $z_0 = 1 + j$ is the bias input, where $j = \sqrt{-1}$ is an imaginary unity. The weight update rules for various parameters of a considered feedforward network of the ***CRSP*** neuron are given here:

Let V_m^{π} be the net potential of mth RSP neuron in the hidden layer then from Eq. (4.1)

$$\begin{aligned} V_m^{\pi} &= \lambda_m W_m^S Z^T + \gamma_m \exp\left(-\left\|Z - W_m^{RB}\right\|^2\right) \\ &\quad + \lambda_m W_m^S Z^T \gamma_m \exp\left(-\left\|Z - W_m^{RB}\right\|^2\right) + w_{0m} z_0 \end{aligned} \quad (4.2)$$

This net internal potential of RSP neuron may also be expressed term wise as follows:

$$V_m^{\pi} = V_m^{\pi 1} + V_m^{\pi 2} + V_m^{\pi 1} V_m^{\pi 2} + w_{0m} z_0 \quad (4.3)$$

From Eq. (3.3), the output of a neuron in the hidden layer can be expressed as

$$Y_m = f\left(\Re\left(V_m^{\pi}\right)\right) + j\ f\left(\Im\left(V_m^{\pi}\right)\right) \tag{4.4}$$

Output of a neuron in the output layer ($n = 1 \ldots N$) of considered network is

$$Y_n = f_C\left(V_n\right) = f_C\left(\sum_{m=1}^{M} w_{mn} Y_m + w_{0n}\ z_0\right) \tag{4.5}$$

Update equation for learning parameters in output layer

$$\Delta w_{mn} = \eta\ \overline{Y}_m\ (\Re(e_n)\ f'(\Re(V_n)) + j\ \Im(e_n)\ f'(\Im(V_n))) \tag{4.6}$$

$$\Delta w_{0n} = \eta\ \overline{z}_0\ (\Re(e_n)\ f'(\Re(V_n)) + j\ \Im(e_n)\ f'(\Im(V_n))) \tag{4.7}$$

The update equations for learning parameters between input and hidden layer are as follows:

$$\Delta w_{lm}^{S} = \frac{\eta}{N}\ \overline{z_l \lambda_m}\left(1 + \overline{V_m^{\pi 2}}\right)\left(\Gamma_m^{\pi} + j\ \xi_m^{\pi}\right) \tag{4.8}$$

$$\begin{aligned}\Delta w_{lm}^{RB} &= \frac{2\eta}{N}\ \exp\left(-\left\|Z - W_m^{RB}\right\|^2\right)\left(z_l - w_{lm}^{RB}\right)\\ &\left[\Gamma_m^{\pi}\left\{\Re\left(\gamma_m\right)\left(1 + \Re\left(V_m^{\pi 1}\right)\right) - \Im\left(\gamma_m\right)\Im\left(V_m^{\pi 1}\right)\right\}\right.\\ &\left.+\xi_m^{\pi}\left\{\Im\left(\gamma_m\right)\left(1 + \Re\left(V_m^{\pi 1}\right)\right) + \Re\left(\gamma_m\right)\Im\left(V_m^{\pi 1}\right)\right\}\right]\end{aligned} \tag{4.9}$$

$$\Delta\lambda_m = \frac{\eta}{N}\overline{\left(W_m^{S}\ Z^{T}\right)}\left(1 + \overline{V_m^{\pi 2}}\right)\left(\Gamma_m^{\pi} + j\ \xi_m^{\pi}\right) \tag{4.10}$$

$$\Delta\gamma_m = \frac{\eta}{N}\ \exp\left(-\left\|Z - W_m^{RB}\right\|^2\right)\left(1 + \overline{V_m^{\pi 1}}\right)\left(\Gamma_m^{\pi} + j\ \xi_m^{\pi}\right) \tag{4.11}$$

$$\Delta w_{0m} = \frac{\eta}{N}\ \overline{z_0}\left(\Gamma_m^{\pi} + j\ \xi_m^{\pi}\right) \tag{4.12}$$

$$\begin{aligned}\text{where } \Gamma_m^{\pi} &= f'(\Re(V_m^{\pi}))\sum_{n=1}^{N}\{\Re(e_n)\ f'(\Re(V_n))\ \Re(w_{mn})\\ &+\Im(e_n)\ f'(\Im(V_n))\ \Im(w_{mn})\}\\ \text{and } \xi_m^{\pi} &= f'(\Im(V_m^{\pi}))\sum_{n=1}^{N}\{\Im(e_n)\ f'(\Im(V_n))\ \Re(w_{mn})\\ &-\Re(e_n)\ f'(\Re(V_n))\ \Im(w_{mn})\}\end{aligned}$$

4.3.4 Model-2

Another neuron architecture for complex-valued signals has been defined by an aggregation function which is also a functional of input signals. Before integration, the signals are preprocessed through summation and radial basis operations. Such processed information is integrated in desired proportion ($\gamma : \lambda$), linearly ($\bigoplus$) in the RSS (Rbf-Sigma-Sigma) model. The compensatory parameters γ and λ specify the contribution of radial basis and summation subfunctions to take into account the vagueness involved. The union or algebraic sum are inserted in the formulas for modeling linear operation, expressed here as $a \oplus b = 1 + a + b$. The novel neuron constructed with this operation is named as RSS. The net potential of this neuron is defined by following aggregation function:

$$\Omega_m(z_1, z_2 \ldots z_L) = \left[\lambda_m \times W_m^S\, Z^T \bigoplus \gamma_m \times \exp\left(-\left\|Z - W_m^{RB}\right\|^2\right)\right] \tag{4.13}$$

The output of this neuron may be expressed as: $Y_m = f_C(\Omega_m(z_1, z_2 \ldots z_L))$

4.3.5 Learning Rules for Model-2

Consider a three-layer network (L-M-N) based on RSS neuron, similar to network of RSP neuron. All weights, bias, and input-output signals are complex numbers. Let $Z = [z_1, z_2 \ldots z_L]$ be the vector of input signals and Z^T is transpose of vector Z. $W_m^S = \left[w_{1m}^S, w_{2m}^S \ldots w_{Lm}^S\right]$ is a vector of weights from inputs ($l = 1 \ldots L$) to summation part of mth RSP neuron and $W_m^{RB} = \left[w_{1m}^{RB}, w_{2m}^{RB} \ldots w_{Lm}^{RB}\right]$ is a vector of weights from inputs to radial basis part of mth RSS neuron. w_0 is a bias and $z_0 = 1 + j$ is the bias input. The error gradient and weight update rules for various weights of a feedforward network of the proposed neuron can be derived as follows:

Let V_m^σ be net potential of mth RSS neuron in the hidden layer then from Eq. (4.13)

$$V_m^\sigma = \lambda_m\, W_m^S\, Z^T + \gamma_m \exp\left(-\left\|Z - W_m^{RB}\right\|^2\right) + w_{0m} z_0 \tag{4.14}$$

From Eq. (3.3), the output of a neuron in the hidden layer can be expressed as

$$Y_m = f\left(\Re\left(V_m^\sigma\right)\right) + jf\left(\Im\left(V_m^\sigma\right)\right) \tag{4.15}$$

Output of a neuron in the output layer ($n = 1 \ldots N$) of considered network is

$$Y_n = f_C(V_n) = f_C\left(\sum_{m=1}^{M} w_{mn} Y_m + w_{0n}\, z_0\right) \tag{4.16}$$

Update equation for parameters in output layer

$$\Delta w_{mn} = \eta\, \overline{Y}_m\, (\Re(e_n)\, f'(\Re(V_n)) + j\, \Im(e_n)\, f'(\Im(V_n))) \tag{4.17}$$

$$\Delta w_{0n} = \eta\, \overline{z}_0\, (\Re(e_n)\, f'(\Re(V_n)) + j\, \Im(e_n)\, f'(\Im(V_n))) \tag{4.18}$$

The update equations for parameters between input and hidden layer are as follows:

$$\Delta w_{lm}^{S} = \frac{\eta}{N}\, \overline{z_l}\, \overline{\lambda}_m \left(\Gamma_m^{\sigma} + j\xi_m^{\sigma}\right) \tag{4.19}$$

$$\Delta w_{lm}^{RB} = \frac{2\eta}{N} \exp\left(-\left\| \mathrm{Z} - \mathrm{W}_m^{RB} \right\|^2\right) \left(z_l - w_{lm}^{RB}\right) \left\{\Gamma_m^{\sigma} \Re(\gamma_m) + \xi_m^{\sigma}\, \Im(\gamma_m)\right\} \tag{4.20}$$

$$\Delta \lambda_m = \frac{\eta}{N}\, \overline{\left(\mathrm{W}_m^{S}\, \mathrm{Z}^T\right)} \left(\Gamma_m^{\sigma} + j\, \xi_m^{\sigma}\right) \tag{4.21}$$

$$\Delta \gamma_m = \frac{\eta}{N} \exp\left(-\left\| \mathrm{Z} - \mathrm{W}_m^{RB} \right\|^2\right) \left(\Gamma_m^{\sigma} + j\, \xi_m^{\sigma}\right) \tag{4.22}$$

$$\Delta w_{0m} = \frac{\eta}{N}\, \overline{z_0} \left(\Gamma_m^{\sigma} + j\, \xi_m^{\sigma}\right) \tag{4.23}$$

where $\Gamma_m^{\sigma} = f'(\Re(V_m^{\sigma})) \sum_{n=1}^{N} \{\Re(e_n)\, f'(\Re(V_n))\, \Re(w_{mn}) + \Im(e_n)\, f'(\Im(V_n))\, \Im(w_{mn})\}$

and $\xi_m^{\sigma} = f'(\Im(V_m^{\sigma})) \sum_{n=1}^{N} \{\Im(e_n)\, f'(\Im(V_n))\, \Re(w_{mn}) - \Re(e_n)\, f'(\Re(V_n))\, \Im(w_{mn})\}$

4.3.6 Model-3

Over the years, a substantial body of evidence has grown to support the presence of nonlinear aggregation of synaptic inputs in the neuron cells [1, 2]. This section exploit a novel fuzzy oriented averaging concept, appeared in [35], to design a very general nonlinear aggregation function whose special cases appear in various types of existing neural aggregation functions. In general, an aggregation operation can be viewed as averaging operation [38]. A brief survey into the history of averaging operations brings out the fact that in 1930 Kolmogoroff [39] and Nagumo [40] acknowledged the family of quasi-arithmetic means as a most general averaging operations. This family has been defined as follows:

$$M_{\phi}(x_1, x_2 \ldots x_n; \omega_1, \omega_2 \ldots \omega_n) = \phi^{-1}\left(\sum_{k=0}^{n} \omega_k\, \phi(x_k)\right) \tag{4.24}$$

where ϕ: $R^n \longrightarrow R$ is a continuous strictly monotonic function and ϕ^{-1} is its inverse function. The function ϕ is called a generator of the quasi-arithmetic mean M_ϕ. $\omega \in [0, 1]^n$ and $\sum_{k=0}^{n} \omega_k = 1$. The class of all quasi-arithmetic means are characterized by function ϕ. One of the very notable class is root-power mean or generalized mean that covers the entire interval between the min and max operations. It is defined corresponding to the function ϕ: $x \longrightarrow x^d$, $d \in R \backslash \{0\}$. The weighted root-power mean (M_d), is defined as:

$$M(x_1, x_2 \ldots x_n; \omega_1, \omega_2 \ldots \omega_n; d) = \left(\sum_{k=0}^{n} \omega_k \, x_k^d \right)^{1/d} \tag{4.25}$$

Dyckhoff and Pedrycz [36] discuss generalized mean as a model of compensative operator that fits the data relatively better. Here, the modifiable degree of compensation is accomplished by changing the value of generalization parameter d. Depending on its value, the model embraces a full spectrum of classical means. In the limit cases $d \to \pm\infty$, model behaves as the maximum and minimum operator respectively. As $d \to 0$, M converges to the geometric mean. Similarly, when $d = -1, 1, 2$ then arguments combined yield their harmonic, arithmetic, and quadratic means, respectively. If we use the generalized mean for aggregation, it is possible to go through all possible variations of means of the input signals to a neuron [35, 36, 38]. This motivated to utilize the idea underlying the weighted root-power mean in Eq. (4.25) to define a new aggregation function for nonconventional neural units in complex domain. The net potential of this complex root-power mean neuron (***C***RPN) may conveniently be expressed as:

$$\Omega(z_1, z_2 \ldots z_n; w_1, w_2 \ldots w_n; d) = \left(\sum_{k=0}^{n} w_k \, z_k^d \right)^{1/d} \tag{4.26}$$

The weighted root-power mean aggregation operation designed a fundamental class of higher-order neuron unit. In comparison to conventional neuron, it gives more freedom to change the functionality of a neuron by choosing the appropriate value of generalization parameter 'd'.

Now, from Eq. 4.26, the output of proposed ***C***RPN may be given as:

$$Y(z_1, z_2 \ldots z_n; w_1, w_2 \ldots w_n; d) = fc\left(\left(\sum_{k=0}^{n} w_k \, z_k^d \right)^{1/d} \right) \tag{4.27}$$

The motivation for using (4.27) is that it gives more freedom to change the functionality of a neuron by choosing the appropriate value of power coefficient d. It is worth indicating that the (4.27) presenting ***C***RPN is general enough and different existing neuron models can be obtained as its special cases by varying the value of d. These are as follows:

When $d = 1$ the **C**RPN model turn out to be conventional neuron (**C**MLP), then from (4.27)

$$Y(z_1, z_2 \ldots z_n; w_1, w_2 \ldots w_n) = fc\left(\sum_{k=0}^{n} w_k \, z_k\right) \tag{4.28}$$

which is the conventional neuron[2] model proposed in real and complex domain as cases apply.

Now, considering all variables in real domain, when $d \to 0$ then (4.25) and (4.27) yield:

$$Y(x_1, x_2 \ldots x_n; w_1, w_2 \ldots w_n) = \lim_{d \to 0} M = f\left(\prod_{k=0}^{n} x_k^{w_k}\right) \tag{4.29}$$

which is a multiplicative neuron proposed in [41] whose functional capability is proved there in.

When $d = -1$ then Eq. 4.27 yields:

$$Y(x_1, x_2 \ldots x_n; w_1, w_2 \ldots w_n) = f\left(\frac{1}{\sum_{k=0}^{n} \frac{w_k}{x_k}}\right) \tag{4.30}$$

which is a harmonic neuron model proposed in [42].

When $d = 2$ then from Eq. 4.27:

$$Y(x_1, x_2 \ldots x_n; w_1, w_2 \ldots w_n) = f\left(\sum_{k=0}^{n} w_k \, x_k^2\right)^{1/2} \tag{4.31}$$

This is a neuron model which is conceptually similar to the quadratic neuron presented in [20, 43].

4.3.7 Learning Rules for Model-3

Consider, a three-layer network (L-M-N) of **C**RPN model, first layer consists of L inputs ($l = 1 \ldots L$), second and output layer has M ($m = 1 \ldots M$) and N

[2] In this book, a neuron with only summation aggregation function is referred to as a 'conventional neuron' and a standard feedforward neural network with these neurons is referred to as as 'MLP' (multilayer perceptron).

$(n = 1 \ldots N)$ proposed neurons, respectively. All weights, bias and input-output signals are complex numbers. Let $Z = [z_1, z_2 \ldots z_L]$ be the vector of input signals. Let $W_m = [w_{1m}, w_{2m} \ldots w_{Lm}]$ be the vector of weights between input layer to m^{th} hidden neuron, while $W_n = [w_{1n}, w_{2n} \ldots w_{Mn}]$ be the vector of weights between hidden layer to nth output neuron. w_0 is the bias weight and z_0 is the bias input. From Eqs. (3.3) and Eq.(4.26), the output of any neuron in the hidden layer can be expressed as:

$$Y_m = f(\Re(V_m)) + j \times f(\Im(V_m)) \quad \text{where} \quad V_m = \left(\sum_{l=0}^{L} w_{lm} \, z_l^d\right)^{1/d} \tag{4.32}$$

Similarly, the output of each neuron in output layer can be expressed as:

$$Y_n = f(\Re(V_n)) + j \times f(\Im(V_n)) \quad \text{where} \quad V_n = \left(\sum_{m=0}^{M} w_{mn} \, Y_m^d\right)^{1/d} \tag{4.33}$$

Update equation for learning parameters in output layer:

$$\Delta w_{mn} = \frac{\eta}{d} \left(\overline{V}_n\right)^{(1-d)} \left(\overline{Y}_m\right)^d \left(\Re(e_n)\, f^{'}(\Re(V_n) + j \times \Im(e_n)\, f^{'}(\Im(V_n))\right) \tag{4.34}$$

The update equation for learning parameters between input and hidden layer is as follows. Let AT_{mn} and VT_{mn} are common terms,

$$AT_{mn} = (A1_{mn} A2_{mn} + A3_{mn} A4_{mn}) \qquad VT_{mn} = (A3_{mn} A2_{mn} - A1_{mn} A4_{mn})$$

$$A1_{mn} = \Re\left(Y_m^d\right) \Re(Y_m) + \Im\left(Y_m^d\right) \Im(Y_m) \quad A2_{mn} = \Re\left(V_n^d\right) \Re(V_n) + \Im\left(V_n^d\right) \Im(V_n)$$

$$A3_{mn} = \Re\left(Y_m^d\right) \Im(Y_m) - \Im\left(Y_m^d\right) \Re(Y_m) \quad A4_{mn} = \Re\left(V_n^d\right) \Im(V_n) - \Im\left(V_n^d\right) \Re(V_n)$$

$$\begin{aligned}
\text{and } \Upsilon_{lm} = \sum_n \frac{1}{\left|V_n^d\right|^2} \Big[& \left\{\Re(e_n)\, f^{'}(\Re(V_n))\, \Re(w_{mn}) + \Im(e_n)\, f^{'}(\Im(V_n))\, \Im(w_{mn})\right\} \\
& \times \left(f^{'}(\Re(V_m))\, AT_{mn} + j \times f^{'}(\Im(V_m))\, VT_{mn}\right) \\
& + j\left\{\Im(e_n)\, f^{'}(\Im(V_n))\, \Re(w_{mn}) - \Re(e_n)\, f^{'}(\Re(V_n))\, \Im(w_{mn})\right\} \\
& \times \left(f^{'}(\Im(V_m))\, AT_{mn} + j \times f^{'}(\Re(V_m))\, VT_{mn}\right)\Big]
\end{aligned} \tag{4.35}$$

$$\Delta w_{lm} = \frac{\eta}{N\, d\, |Y_m|^2} \left(\overline{V}_m\right)^{(1-d)} \left(\overline{z}_l\right)^d \, \Upsilon_{lm} \tag{4.36}$$

4.4 Performance Variation Among Conventional and Higher-Order Neurons

In order to estimate the strength and effectiveness of a neural network of considered higher-order neurons with learning algorithms in real and complex domain, this section presents performance evaluation on different types of problems in real and complex domain. In case of real-valued problems, the input-output data is assigned to real part and the value very close to zero is assigned to imaginary part, to get the complex-valued data [25, 34]. The results obtained are compared with the ones shown by other neural networks and training algorithms. An intelligent choice of RPROP parameters and weight initialization gives good results. The computational power and approximation capability have been compared in terms of number of epochs, learning parameters, network topology, and testing error along with other statistical performance evaluation metrics like correlation, error variance, and Akaikes information criteria (AIC) [44]. AIC evaluates the goodness of fit of model based on mean square error for training data and number of estimated parameters. For analysis purposes, the number of learning parameters in each complex-valued weight is counted as two [32].

4.4.1 Real Domain Problems

4.4.1.1 The Wine Recognition Data Problem

The wine dataset is a result of chemical analysis of wine with 13 constituents found in each of the three types of wines. Dataset in [45] provide 178 instances of all three classes. In our experiments some random 60 % data are taken for training and rest 40 % for testing. The performance result using different type of networks with four training algorithms viz ***RBP***, ***CBP*** (with $\eta = 0.003$) and ***RRPROP***, ***CRPROP*** (with $\mu^- = 0.5, \mu^+ = 1.2, \Delta_{\min} = 10^{(-6)}, \Delta_{\max} = 0.005, \Delta_0 = 0.1$) is presented in Table 4.1. A very poor convergence is observed with backpropagation algorithm. The number of training epochs is drastically reduced when proposed ***CRPROP*** algorithm is used for training. The significant advantage of using ***CRPROP*** algorithm is that a perfect result (no misclassification) is achieved with least training epochs. Results reveals the superiority of the different neurons in a complex domain in all respects.

4.4.1.2 Ionosphere Data

This radar data was collected by a system in Goose Bay, Labrador, Johns Hopkins University and available in the database [45]. These radar returns belongs to two classes, "Good" radar returns are those showing evidence of some type of structure in the ionosphere and "Bad" returns are those that do not, their signals pass through

Table 4.1 Comparison of training and testing performance for Wine Data problem

S. No.	Training algorithm	Neuron type	Network	Average epochs	MSE training	Testing error (%)
1	***R*BP**	***R*-MLP**	13-6-1	45,000	0.0065	2.8
		***R*RPN** ($d = 0.85$)	13-4-1	35,000	0.0034	2.8
		***R*RSS**	13-2-1	40,000	0.0039	2.8
		***R*RSP**	13-2-1	29,000	0.0040	2.8
2	***R*RPROP**	***R*MLP**	13-5-1	2,000	0.0047	2.8
		***R*RPN** ($d = 0.85$)	13-4-1	2,000	0.0019	1.4
		***R*RSS**	13-2-1	2,000	0.0032	2.8
		***R*RSP**	13-2-1	2,000	0.0028	1.4
3	CBP	CMLP	13-3-1	35,000	0.0034	1.4
		CRPN ($d = 0.85$)	13-1-1	17,000	0.0026	1.4
		CRSS	13-1	20,000	0.0039	2.8
		CRSP	13-1	15,000	0.0029	1.4
4	CRPROP	CMLP	13-1-1	2,000	0.0012	0.0
		CRPN ($d = 1.1$)	13-1-1	1,500	0.0011	0.0
		CRSS	13-1	1,500	0.0027	1.4
		CRSP	13-1	1,500	0.0020	0.0

the ionosphere. The targets were free electrons in the ionosphere. Simulations use first 200 instances for training and remaining 150 instances for testing. Table 4.2 presents comparative performance of different networks with training algorithms viz ***RBP***, ***CBP*** ($\eta = 0.004$) and ***RRPROP***, ***CRPROP*** ($\mu^- = 0.5, \mu^+ = 1.2, \Delta_{min} = 10^{(-6)}, \Delta_{max} = 0.003, \Delta_0 = 0.1$). Results clearly mention that a RPN-based network achieve minimum test error with least training epochs when implemented in a complex domain.

4.4.2 Complex Domain Problems

4.4.2.1 Approximation of Vector Operations

A complex number corresponds to a vector with the coordinates as its components. In fact a complex number 'z' is referred as a radius vector from the origin to the point that represents 'z'. The vector interpretation of complex number is very useful in analyzing various geometrical relationships on a plane. The set of complex number is well equipped with basic algebraic operations. The summation (SUM), subtraction (SUB), multiplication (MULT), and Quotient (QUOT) of two complex number is a complex number and is very helpful in defining different vector operations in the two dimensional space. This experiment is conducted to design a learning machine, which

Table 4.2 Comparison of training and testing performance for Ionosphere Data problem

S. No.	Training algorithm	Neuron type	Network	Average epochs	MSE training	Testing error (%)
1	***R*BP**	***R*MLP**	34-3-1	8,000	0.0068	6.7
		***R*RPN ($d = 0.85$)**	34-3-1	8,000	0.0092	6.0
		***R*RSS**	34-1-1	14,000	0.0095	7.3
		***R*RSP**	34-1-1	14,000	0.0090	6.0
2	***R*RPROP**	***R*MLP**	34-3-1	2,000	0.0094	7.3
		***R*RPN ($d = 0.85$)**	34-3-1	2,000	0.0083	6.0
		***R*RSS**	34-1-1	3,000	0.013	8.4
		***R*RSP**	34-1-1	3,000	0.012	8.0
3	***C*BP**	***C*MLP**	34-2-1	7,000	0.0065	6.0
		***C*RPN ($d = 0.85$)**	34-1-1	7,000	0.0070	4.0
		***C*RSS**	34-1	14,000	0.016	7.3
		***C*RSP**	34-1	14,000	0.0095	6.0
4	***C*RPROP**	***C*MLP**	34-2-1	1,500	0.001	6.0
		***C*RPN ($d = 0.85$)**	34-1-1	1,500	0.007	4.0
		***C*RSS**	34-1	3,000	0.013	8.0
		***C*RSP**	34-1	3,000	0.013	7.3

can approximate above four vector operations through a single network. Let $z1 = r1e^{j\vartheta 1}$ and $z2 = r2e^{j\vartheta 2}$, where $-pi \leq \vartheta 1, \vartheta 2 \leq pi$ and $0.1 \leq r \leq 0.5$. A set of 200 patterns was randomly chosen for training and approximation of trained network was tested on other 1000 patterns. Table 4.3 analyzes the performance of different neuron-based networks with two training algorithms viz ***C*BP** ($\eta = 0.0005$) and ***C*RPROP** ($\mu^- = 0.5, \mu^+ = 1.2, \Delta_{\min} = 10^{(-6)}, \Delta_{\max} = 0.001, \Delta_0 = 0.005$) on above test data. Results reported in Table 4.3 are from reasonably smaller network topology which can yield fairly good accuracy. On increasing the number of neurons in a hidden layer, performance measure in SUM and SUB operations are unaffected while slow improvement is observed in MULT and QUOT operations. ***C*RSP** neuron-based network outperforms, especially in approximation of MULT and QUOT operations with quiet lesser number of learning parameters.

4.4.2.2 2D Gabor Function

The 2D Gabor function is a unique function that achieve the lower bound for the space-frequency uncertainty product. It is a measure of a function's simultaneous localization in both spatial and frequency domains. It is well known that the highly oriented simple cell receptive fields in the visual cortex of mammals can be closely modeled by 2D Gabor functions [46], which are Gaussians modulated by sinusoidal

Table 4.3 Training and testing performance for approximation of vector operations

		CMLP		CRPN ($d = 1.05$)		CRSS		CRSP	
Algorithm		CBP	CRPROP	CBP	CRPROP	CBP	CRPROP	CBP	CRPROP
Network		2-18-4	2-18-4	2-18-4	2-18-4	2-10-4	2-10-4	2-10-4	2-10-4
Parameters		130	130	130	130	114	114	114	114
MSE (training)		0.0010	0.0015	0.0014	0.00057	0.0021	0.0029	0.0013	0.0021
MSE (testing)	SUM	0.0008	0.0007	0.0022	0.0020	0.0009	0.0007	0.0006	0.0005
	SUB	0.0009	0.0010	0.0026	0.0020	0.0018	0.0009	0.0010	0.0007
	MULT	0.0098	0.0010	0.0072	0.0014	0.0092	0.0004	0.0033	0.0003
	QUOT	0.0029	0.0074	0.0022	0.0010	0.0063	0.0008	0.0013	0.0006
Correlation	SUM	0.9983	0.9983	0.9796	0.8838	0.9954	0.9876	0.9975	0.9984
	SUB	0.9986	0.9963	0.9786	0.9858	0.9958	0.9890	0.9971	0.9980
	MULT	0.1296	0.9063	0.1886	0.8758	0.1454	0.9376	0.1748	0.9812
	QUOT	0.1096	0.1463	0.0241	0.0218	0.0554	0.1196	0.2780	0.3179
Error variance	SUM	0.0009	0.0007	0.0021	0.0022	0.0010	0.0007	0.0006	0.0005
	SUB	0.0009	0.0009	0.0022	0.0020	0.0013	0.0009	0.0010	0.0007
	MULT	0.0094	0.0009	0.0072	0.0014	0.0082	0.0035	0.0030	0.0003
	QUOT	0.0041	0.0084	0.0044	0.0021	0.0043	0.0022	0.0013	0.0015
AIC		−6.14	−6.13	−6.22	−6.18	−6.07	−5.58	−6.47	−6.86
Average epochs		10,000	2,000	10,000	2,000	10,000	2,000	10,000	2,000

functions. The convolution version of complex 2D Gabor functions have the following form:

$$g(x_1, x_2) = \frac{1}{2\pi\lambda\sigma^2} e^{-\left\{\left[(x_1/\lambda)^2 + x_2^2\right]/2\sigma^2\right\}} e^{2\pi j(u_0 x_1 + v_0 x_2)} \tag{4.37}$$

Here, λ is the aspect ratio, σ is the scale factor, and (u_0, v_0) are modulation parameters. When $\lambda = 1$, the $g(x_1, x_2)$ becomes circularly symmetric. Thus, the Gabor function to be approximated is

$$g(x_1, x_2) = \frac{1}{2\pi(0.5)^2} e^{-[(x_1^2 + x_2^2)/2(0.5)^2]} e^{2\pi j(x_1 + x_2)} \tag{4.38}$$

Shin and Ghosh [47] and Li [48], considered either imaginary or real part of above function in their simulations. But this section uses complete 2D Gabor function for approximation. For training, 100 input points were randomly selected from an evenly spaced 10×10 grid on $-0.5 \le x_1, x_2 \le 0.5$. Similarly, 900 input points were randomly selected for testing from an evenly spaced 30×30 grid on $-0.5 \le x_1, x_2 \le 0.5$. Figure 4.2 presents the real and imaginary part of 2D Gabor function for this

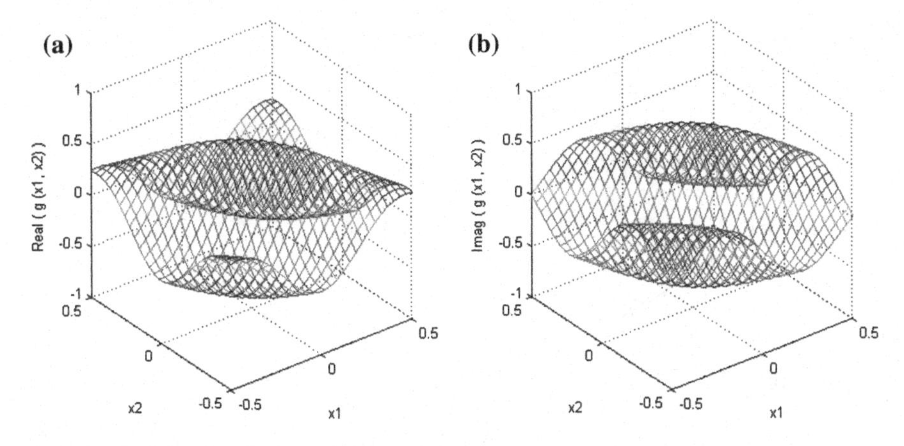

Fig. 4.2 2D Gabor function; **a** real part and **b** imaginary part

Table 4.4 Comparison of training and testing performance for 2D Gabor Function

	***C*MLP**		***C*RPN ($d = 0.96$)**		***C*RSS**		***C*RSP**	
Algorithm	***C*BP**	***C*RPROP**	***C*BP**	***C*RPROP**	***C*BP**	***C*RPROP**	***C*BP**	***C*RPROP**
Network	2-10-1	2-10-1	2-10-1	2-10-1	2-5-1	2-5-1	2-5-1	2-5-1
Parameters	41	41	41	41	41	41	41	41
MSE (training)	0.0018	0.00088	0.0026	0.00091	0.001	0.00097	0.00069	0.0004
MSE (testing)	0.0039	0.0016	0.0048	0.0017	0.0019	0.00099	0.0011	0.00062
Correlation	0.9816	0.9913	0.9886	0.9958	0.9954	0.9976	0.9978	0.9979
Error variance	0.0039	0.0021	0.0046	0.0017	0.0019	0.00079	0.0011	0.00063
AIC	−6.18	−6.73	−5.78	−6.82	−6.67	−7.32	−7.27	−7.76
Average epochs	9,000	2,000	9,000	2,000	9,000	2,000	9,000	2,000

test set. For comparison, we tried the approximation of above function with different networks using two training algorithms viz ***C*BP** ($\eta = 0.01$) and ***C*RPROP** $\left(\mu^- = 0.5, \mu^+ = 1.2, \Delta_{\min} = 10^{(-6)}, \Delta_{\max} = 0.1, \Delta_0 = 0.1\right)$. After experimenting with different numbers of hidden neurons and training epochs, the best result is reported in Table 4.4. In order to graphically visualize the 2-D Gabor function, the real and imaginary part of different network's outputs are shown in Fig. 4.3. Results in Table 4.4 clearly demonstrate that the ***C*RSP** network with ***C*RPROP** perform best with least error variance and testing error in only 2000 epochs, keeping same number of learning parameters.

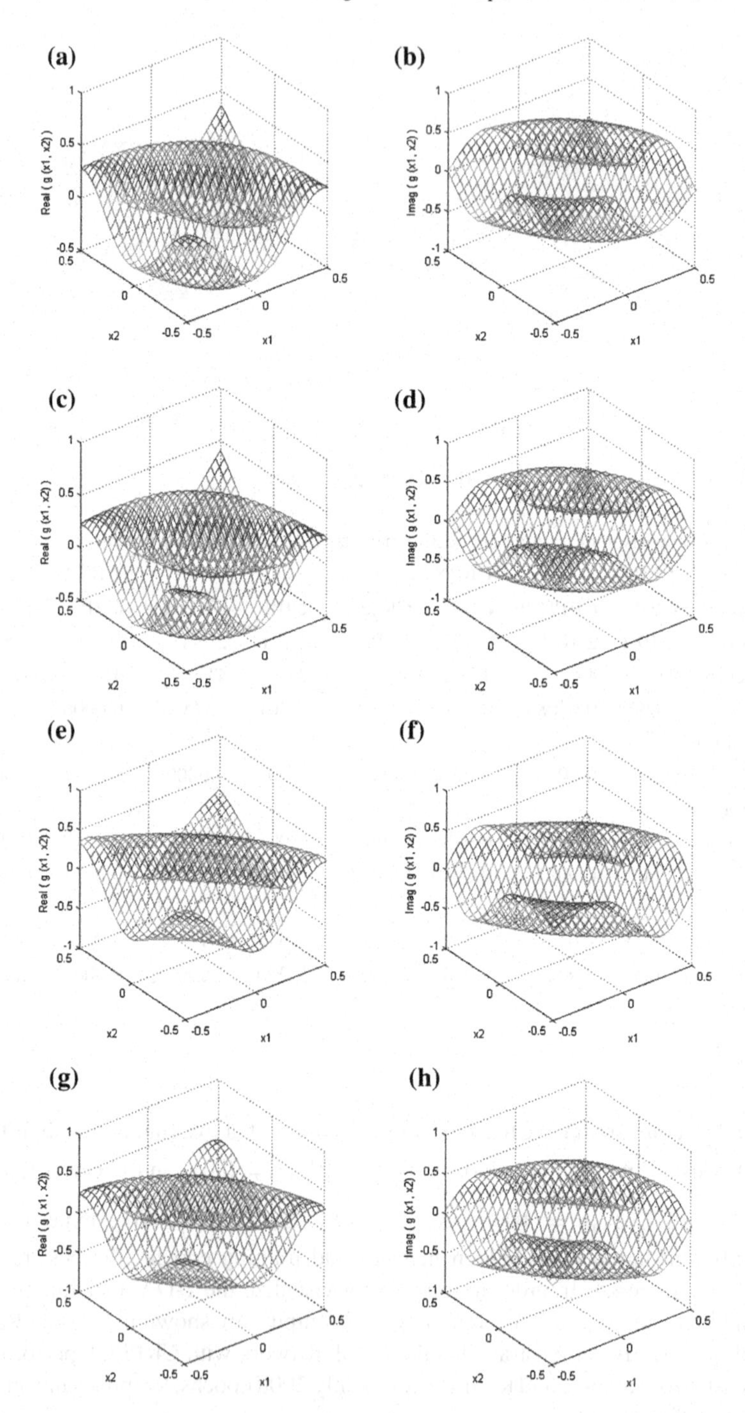

Fig. 4.3 Approximation of 2D Gabor function with *C*MLP (**a** real and **b** imaginary part), *C*RPN (**c** real and **d** imaginary part), *C*RSS (**e** real and **f** imaginary part) and *C*RSP (**g** real and **h** imaginary part)

4.5 Concluding Remarks

It is well known that the conventional real-valued neuron takes large number of neurons in a network, which indeed increases the complexity of a network, when used to solve any single- or high- dimensional problem. The higher the complexity of a network, greater will be its computational complexity, time, and memory requirement. The complexity of ANN can only be reduced by using less number of neurons or by using neurons which can process high-dimensional data as a single quantity. In case of real-valued problems, the complexity of ANN can also be effectively reduced by implementing through complex-valued neural network. The trade-off between considering complex domain implementation and higher-order neural network lead to lesser number of learning cycles, better class distinctiveness and superior mapping accuracy in simulation results. The conventional neuron in a MLP or $\mathbb{C}$MLP has a linear correlation among input signals; thus such a neuron model, when used to solve the problems, always appear inferior in all respects to achieve performance similar to higher-order neurons. The fact that the number of unknowns (learning weights) to be determined in such a network grows with the number of neurons and hidden layers. Which in turn lead to quiet slow processing of neural network.

Neural networks today are much more than just the simple network of conventional neurons. The new insights available from neuroscience have presented nonlinear neuronal activities in a cell body. This motivates one to investigate the feasibility of constructing nonlinear aggregation functions, which will serve as a basis for construction of powerful neuron models. The various researchers have described the power and other advantages of higher-order neuron with respect to the conventional neurons. The computational power of the neuron depends on its order, a higher-order neuron can posses better mapping and classification capabilities. However, with an increase in number of terms in the polynomial expression for the higher-order neuron, it is exceedingly difficult to train a network of such neurons. Considering this basic drawback of higher-order neuron, this chapter presented three efficient neuron models for science and engineering applications. Unlike the other higher-order neurons, these models are simpler in terms of its parameters and do not need to determine the monomial structure prior to training of the neuron model. The weight update rules using backpropagation learning algorithm is provided for the feedforward neural networks for presented models. These models can serve as universal approximators and can be conveniently used with conventional or other neuron in a network. The computing and generalization capabilities of these neurons have been further provided in Chaps. 5 and 7 which will better demonstrate the motivation of this chapter.

References

1. Koch, C., Poggio, T.: Multiplying with synapses and neurons. In: McKenna, T., Davis, J., Zornetzer, S.F. (eds.) Single Neuron Computation, pp. 315–345. Academic, Boston, MA (1992)
2. Mel, B.W.: Information processing in dendritic trees. Neural Comput. **6**, 1031–1085 (1995)

3. Arcas, B.A., Fairhall, A.L., Bialek, W.: What can a single neuron compute? In: Leen, T., Dietterich, T., Tresp, V. (eds.) Advances in Neural Information Processing, pp. 75–81. MIT press, Cambridge (2001)
4. McCulloch, W.S., Pitts, W.: A logical calculation of the ideas immanent in nervous activity. Bull. Math. Biophys. **5**, 115–133 (1943)
5. Koch, C.: Biophysics of Computation: Information Processing in Single Neurons. Oxford University Press, New York (1999)
6. Mel, B.W., Koch, C.: Sigma-pi learning : on radial basis functions and cortical associative learning. In: Touretzky, D.S. (ed.) Advances in Neural Information Processing Systems 2, pp. 474–481. Morgan-Kaufmann, San Mateo, CA (1990)
7. Durbin, R., Rumelhart, R.: Product units: a computationally powerful and biologically plausible extension to backpropagation networks. Neural Comput. **1**, 133–142 (1989)
8. Bukovsky, I., Bila, J., Gupta, M.M., Hou, Z.G., Homma, N.: Foundation and classification of nonconventional neural units and paradigm of nonsynaptic neural interaction. In: Wang, Y. (ed.) (University of Calgary, Canada) Discoveries and Breakthroughs in Cognitive Informatics and Natural Intelligence (in the ACINI book series). IGI, Hershey PA, USA (ISBN: 978-1-60566-902-1) (2009)
9. Taylor, J.G., Commbes, S.: Learning higher order correlations. Neural Netw. **6**, 423–428 (1993)
10. Cotter, N.E.: The Stone-Weierstrass theorem and its application to neural networks. IEEE Trans. Neural Netw. **1**, 290–295 (1990)
11. Shin, Y., Ghosh, J.: The Pi-sigma Network: an efficient higher-order neural network for pattern classification and function approximation. Proceedings of the International Joint Conference on Neural Networks, pp. 13–18 (1991)
12. Heywood, M., Noakes, P.: A framework for improved training of Sigma-Pi networks. IEEE Trans. Neural Netw. **6**, 893–903 (1996)
13. Chen, M.S., Manry, M.T.: Conventional modeling of the multilayer perceptron using polynomial basis functions. IEEE Trans. Neural Netw. **4**, 164–166 (1993)
14. Anthony, A., Holden, S.B.: Quantifying generalization in linearly weighted neural networks. Complex Syst. **18**, 91–114 (1994)
15. Chen, S., Billings, S.A.: Neural networks for nonlinear dynamic system modeling and identification. Int. J. Contr. **56**(2), 319–346 (1992)
16. Schmidt, W., Davis, J.: Pattern recognition properties of various feature spaces for higher order neural networks. IEEE Trans. Pattern Anal. Mach. Intell. **15**, 795–801 (1993)
17. Kosmatopoulos, E., Polycarpou, M., Christodoulou, M., Ioannou, P.: High-order neural network structures for identification of dynamical systems. IEEE Trans. Neural Netw. **6**(2), 422–431 (1995)
18. Liu, G.P., Kadirkamanathan, V., Billings, S.A.: On-line identification of nonlinear systems using volterra polynomial basis function neural networks. Neural Netw. **11**(9), 1645–1657 (1998)
19. Elder, J.F., Brown D.E.: Induction and polynomial networks. In: Fraser, M.D. (ed.) Network Models for Control and Processing, pp. 143–198. Intellect Books, Exeter, UK (2000)
20. Bukovsky, I., Redlapalli, S., Gupta, M.M.: Quadratic and cubic neural units for identification and fast state feedback control of unknown non-linear dynamic systems. Fourth International Symposium on Uncertainty Modeling and Analysis ISUMA 2003, pp. 330–334 (ISBN 0-7695-1997-0). IEEE Computer Society, Maryland, USA (2003)
21. Hou, Z.G., Song, K.Y., Gupta, M.M., Tan, M.: Neural units with higher-order synaptic operations for robotic image processing applications. Soft Comput. **11**(3), 221–228 (2007)
22. Nikolaev, N.Y., Iba, H.: Adaptive Learning of Polynomial Networks: Genetic Programming, Backpropagation and Bayesian Methods (ISBN: 0-387-31239-0, series: Genetic and Evolutionary Computation), vol. XIV, p. 316. Springer, New York (2006)
23. Zhang, M. (ed.) (Christopher Newport University): Artificial Higher Order Neural Networks for Economics and Business (ISBN: 978-1-59904-897-0). IGI-Global, Hershey, USA (2008)
24. Rosenblatt, F.: The perceptron : a probabilistic model for information storage and organization in the brain. Psychol. Rev. **65**, 231–237 (1958)

25. Tripathi, B.K., Kalra, P.K.: On efficient learning machine with root power mean neuron in complex domain **22**(5), 727–738, (ISSN: 1045-9227). IEEE Trans. Neural Netw. (2011)
26. Tripathi, B.K., Kalra, P.K.: Complex generalized-mean neuron model and its applications. Appl. Soft Comput. **11**(1), 768–777 (Elsevier Science) (2011)
27. Tripathi, B.K., Kalra, P.K.: Functional mapping with complex higher order compensatory neuron model. World Congress on Computational Intelligence (WCCI-2010), July 18–23. IEEE Xplore, Barcelona, Spain (2010). Proc. IEEE Int. Joint Conf. Neural Netw. **22**(5), 727–738 (ISSN: 1098–7576) (2011)
28. Hirose, A.: Complex-Valued Neural Networks. Springer, New York (2006)
29. Piazza, F., Benvenuto, N.: On the complex backpropagation algorithm. IEEE Trans. Signal Process. **40**(4), 967–969 (1992)
30. Kim, T., Adali, T.: Approximation by fully complex multilayer perceptrons. Neural Comput. **15**, 1641–1666 (2003)
31. Shin, Y., Jin, K.-S., Yoon, B.-Y.: A complex pi-sigma network and its application to equalization of nonlinear satellite channels. In: Proceedings of the IEEE International Conference on Neural Networks (1997)
32. Nitta, T.: An extension of the back-propagation algorithm to complex numbers. Neural Netw. **10**(8), 1391–1415 (1997)
33. Nitta, T.: An analysis of the fundamental structure of complex-valued neurons. Neural Process. Lett. **12**, 239–246 (2000)
34. Tripathi, B.K., Kalra, P.K.: The novel aggregation function based neuron models in complex domain. Soft Comput. **14**(10), 1069–1081 (Springer) (2010)
35. Piegat, A.: Fuzzy Modeling and Control. Springer, Heidelberg (2001)
36. Dyckhoff, H., Pedrycz, W.: Generalized means as model of compensative connectives. Fuzzy Sets Syst. **14**, 143–154 (1984)
37. Lee, C.C., Chung, P.C., Tsai, J.R., Chang, C.I.: Robust radial basis function neural network. IEEE Trans. Syst. Man Cybern. B Cybern. **29**(6), 674–685 (1999)
38. Dubois, D., Prade, H.: A review of fuzzy set aggregation connectives. Inf. Sci. **36**(1–2), 85–121 (1985)
39. Kolmogoroff, A.N.: Sur la notion de la moyenne. Acad. Naz. Lincei Mem. Cl. Sci. Fis. Mat. Natur. Sez. **12**, 388–391 (1930)
40. Nagumo, M.: Uber eine Klasse der Mittelwerte. Japan. J. Math. **7**, 71–79 (1930)
41. Schmitt, M.: On the complexity of computing and learning with multiplicative neural networks. Neural Comput. **14**, 241–301 (2001)
42. Shiblee, Md., Chandra, B., Kalra, P.K.: New neuron model for blind source separation. In: Proceedings of the International Conference on Neural Information Processing, November 25–28 (2008)
43. Georgiou, G.M.: Exact interpolation and learning in quadratic neural networks. In Proceedings IJCNN, Vancouver, BC, Canada, July 16–21 (2006)
44. Foggel, D.B.: An information criterion for optimal neural network selection. IEEE Trans. Neural Netw. **2**(5), 490–497 (1991)
45. Blake, C.L., Merz, C.J.: UCI repository of machine learning database. http://www.ics.uci.edu/mealrn/MLRepository.html. University of California, Department of Information and Computer Science (1998)
46. Daugman, J.G.: Entropy reduction and decorrelation in visual coding by oriented neural receptive fields. IEEE Trans. Biomed. Eng. **36**(1), 107–114 (1989)
47. Shin, Y., Ghosh, J.: Ridge polynomial networks. IEEE Trans. Neural Netw. **6**(3), 610–622 (1995)
48. Li, C.-K.: A sigma-pi-sigma neural network. Neural Process. Lett. **17**, 1–19 (2003)

Chapter 5
High-Dimensional Mapping

Abstract The complex plane is the geometric representation of complex numbers established by the real axis and the orthogonal imaginary axis. A point on the complex plane can be viewed as a complex number with X and Y coordinates regarded as real and imaginary parts of the number. It can be thought of as a modified Cartesian plane, where real part is represented by a displacement along the X-axis and imaginary part by a displacement along the Y-axis. The set of complex numbers is a Field equipped with basic algebraic properties of addition and multiplication operations [1], and hence gives a perfect platform of operation. The properties of the complex plane are different from those of the real line. A complex number have a nonnegative modulus and an argument (Arg) associated with it that locates the complex number uniquely on the plane. It is natural to represent a nonzero complex number with a directed line segment or vector on the complex plane. The extension of traditional real-valued neuron on complex plane has varied its structure from single dimension to two dimensions. Real-valued neuron administers motion on real line, while learning with a complex-valued neuron applies a linear transformation, called 2D motion, [2, 3] to each input signal (complex number on plane). Thus, learning in a complex-valued neural network (CVNN) is characterized with the complex-valued signals flowing through the network, and has ability to capture two dimension patterns naturally. Therefore, the concept of complex plane allows a geometric interpretation of complex numbers in CVNN. The present chapter investigates and explores the mapping properties of the CVNN through some problems of mapping to bring forth the differences between CVNN and ANN, where the stress was on problems that CVNN solves and ANN does not.

5.1 Mapping Properties of Complex-Valued Neural Networks

A function is an association that maps points in a domain to points in the range. Many properties are associated with functional maps—continuity, differentiability, and compactness among many others. In most problems of practical interest, the actual function that governs the input–output behavior is unknown as it becomes increasingly difficult to treat the differential equations that govern the system as

B.K. Tripathi, *High Dimensional Neurocomputing*,
Studies in Computational Intelligence 571, DOI 10.1007/978-81-322-2074-9_5

the number of variables increase or the equation becomes highly nonlinear. However, artificial neural networks can be taught to perform the mapping by treating the dynamical system as a black-box and collecting the input and the corresponding output values, and subjecting the network for training based on these data points. It was observed in researches that initial data points are the most important factor that affects the generalization performance of the neural network design, to other parameters being complexity of the network, dimension of parameters in that order. In practical applications hence, it is the mapping properties of the neural network that are put to use. The validation of the complex-valued neural network (CVNN) against the benchmarks was the first step, while the ability of these networks to map the dynamics of the problems at hand is the step of practical importance.

The mapping properties of the CVNN are the subject of the present chapter. In actual application, the form of the mapping is unknown but error backpropagation algorithm-based neural network captures the mapping [subject to Kolmogorov conditions (Kolmogorov 1957)] through data points. The learning convergence theorem for complex variables (Nitta 1997) is the assurance one needs to establish complex weights exist that solve the mapping problem at hand. To reach to the point in the weights space, the process of training based on gradient descent is employed. The method involves computing the complex gradient and updating the weights based on the slope as obtained from the gradient formula. Though, gradient descent based neural networks have been used extensively to tackle the problems posed by the industry. On the other hand, the CVNN's mapping properties per se have not been studied in literature as yet. Nitta [5] reported some problems of mapping to bring forth the differences between CVNN and ANN; later B K Tripathi (2010) extended mapping problems to more wide spectrum. The present chapter investigates and explores the mapping properties of the CVNN where split-type activation function-based networks are arranged in a performance echelon for each problem studied. The CVNN learns to capture the transformation (magnitude and argument), these networks can be employed to perform the map the vector fields on plane.

5.2 Conformal Mapping on Plane

The complex number is directly related to two-dimensional plane. Variety of mappings or transformations on a plane are used to solve a number of mathematical and practical engineering problems [1, 4]. Such a mapping on complex plane preserves the angles between oriented curves and the phase of each point on the curve is also maintained during transformation. As described in [2, 5], the complex-valued signals flowing through a complex domain network are the unit of learning, which enable to learn 2D motion of signals. In contrast, a neural network in a real domain administers single dimension motion of signals. This is the main reason as to why a neural network extended to a complex domain which can learn mappings on plane, while an equivalent real domain network cannot. Conformal mapping is used to map the complicated regions conformally onto simpler, standard regions, where boundary value problems are easier to solve. One of its novel applications is bilinear transformation, which is expressed as the quotient of two linear expressions, as:

$$\Psi(z) = \frac{(az + b)}{(cz + d)}, \tag{5.1}$$

whose special case is linear transformation, $az + b$; where $a, b, c, d \in \boldsymbol{C}$ are complex constants with a restriction, $ab \neq cd$. Our experiments confirmed that a complex domain neural network has the ability to learn and generalize this transformation with small error.

In [5], author presented the generalization ability of complex MLP for each basic class of transformations, viz. *Scaling, Rotation, and Translation*. The three basic transformations are used to convey dominant geometric characteristics from a complex plane to same or other complex plane. This chapter presents the generalization performance of neurons in Chap. 4 for the composition of two or three basic transformations and a bilinear transformation. Further, elaboration in this chapter make mapping on plane more clear in the context of neural network.

Let input training patterns form a set of points of radius vector $r\, e^{j\vartheta}$. These equal interval points lie on the straight line passing through a reference (mid of line, may be origin), making an angle ϑ with a real axis. The corresponding output patterns form a set of points $(\alpha\ r\ e^{j\,(\vartheta+\tau)}\ +\ \beta\ e^{j\,\sigma})$, which represent the composition of three basic transformations, where $a = \alpha\, e^{j\,\tau}$ and $b = \beta\, e^{j\,\sigma}$. All points in training and testing patterns are within a unit circle centered at origin $(0 \leq r \leq 1)$, and all the angles vary from 0 to $2\,\pi$. Figure 5.1 presents the input–output mapping of a set of sample training patterns over a line.

Definition 5.1 Let $(0 < r, R \leq 1)$ and $(0 \leq \vartheta, \phi\ \leq 2\pi)$. The network is trained for mapping Ψ: $\boldsymbol{C} \rightarrow \boldsymbol{C}$, over a set of points from $r\, e^{j\vartheta}$ to $(\alpha\ r\ e^{j\,(\vartheta+\tau)} + \beta\ e^{j\,\sigma})$. The input test points are set of points $R\, e^{j\phi}$ on the locus of geometric curve then network generates the locus of curve with points $(\alpha\ R\ e^{j\,(\phi+\tau)} + \beta\ e^{j\,\sigma} + [TheError])$. $[TheError]$ is the generalization error, which is a complex number and denotes the difference between an actual output test points and the expected output test points in the transformation. Note, that the network has learned scaling factor α, an angle of rotation τ and a displacement distance β in the direction of σ.

The conformal mapping preserves angles both in magnitude and sense [1], when a problem is mapped from a complex domain D_z to other complex domain D_{zz} in which the solution is sought. In [5], author discussed the generalization error and derived the equation for this error of transformation. He has shown that the error differs with each cycle of learning as it depends on the value of learning parameters. The error increases as the distance between the input test and training points increases and decreases as the test points become closer to the training points. Bilinear transformation is an important class of conformal mapping, and may be considered as benchmark to explore the ability of CVNN in processing two-dimensional information directly. Linear transformation is its special case. In order to validate magnitude and phase preserving property of CVNN and corresponding learning algorithms,

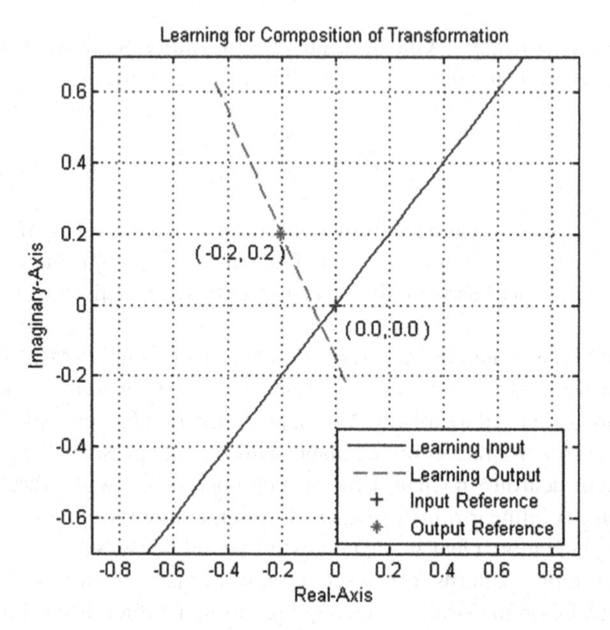

Fig. 5.1 Learning patterns for analysis, mapping shows scaling by factor α, an angle of rotation τ and a displacement by distance β in the direction of σ

various simulations examples have been presented here for learning and generalization of mapping Ψ. In all examples, this chapter has considered a 2-M-2 network, where M is the number of proposed neurons in a hidden layer. A set of points in the first input lie on locus of a input curve and second input is the reference point of input curve. Similarly, the first output neuron gives the locus of transformed curve and second output is its reference point. The transformations of different curves are graphically shown in various figures of this section. *Black* color represents the input test figure, desired output is shown by doted *Blue* color, and actual results are shown in *Red* color. Different examples presented in this section for linear and bilinear transformations are significant in performance analysis of novel neurons in a complex domain.

5.2.1 Linear Transformation

In the present context, the linear mapping is described as a particular case of the conformal mapping (bilinear transformation). If $\Psi(z)$ is an arbitrary bilinear transformation [refer to Eq. (5.1)] and $c = 0$, then $\Psi(z)$ reduces to linear transformation (composition of *Scaling, Rotation, and Translation*) of the form:

$$\Psi 1(z) = a\,z + b \tag{5.2}$$

where $z, a, b \in \boldsymbol{C}$. Evidently, this is an expansion or contraction by a factor $|a|$ and rotation through an angle equal to Arg$a$ in counterclockwise direction, followed by translation in a direction defined by the Arg$b$ through a distance equal to $|b|$. In following two Examples 5.1 and 5.2, different networks (2-M-2) are trained for input–output mapping (refer to Fig. 5.1) over a set of points lying on a line and passing through a reference point. The hidden layer of considered networks contain one $\boldsymbol{C}$RSP or $\boldsymbol{C}$RSS or two $\boldsymbol{C}$RPN or three conventional neurons, respectively. The generalization is tested over other standard geometric curves like circle and ellipse.

Example 5.1 Scaling, Rotation, and Translation
This example investigates the behavior of different networks, which learned the composition of all three transformations defined in Eq. (5.2). All networks are run up to 4,500 epochs with $\boldsymbol{C}$BP ($\eta = 0.001$) and 1,000 epochs with $\boldsymbol{C}$RPROP algorithm ($\mu^- = 0.4, \mu^+ = 1.2, \Delta_{\min} = 10^{(-6)}, \Delta_{\max} = 0.04, \Delta_0 = 0.01$). The learning patterns form a set of points z, which are contracted by factor $\alpha = 1/2$, rotated counterclockwise over $3\pi/4$ radians and displaced by $b = (-0.1 + j \times 0.2)$. There are 21 training inputs lie on the line $y = x$, $(-1/\sqrt{2} \leq x \leq 1/\sqrt{2})$, referenced at origin. The training output points lie on the line $y = 0.2$, $(-0.6 \leq x \leq 0.4)$ with reference $(-0.1, 0.2)$. Transformations in Fig. 5.2 shows the generalization over circle with different networks and learning algorithms. The input test points lie on the circle $x^2 + y^2 = R^2$, with $R = 0.9$. The desired output points should lie on the circle $(x+0.1)^2 + (y-0.2)^2 = (R/2)^2$, where radius vector of each point is rotated by $3\pi/4$. The rotation of the circle is denoted by a small opening.

Example 5.2 Scaling and Rotation
Here, we investigate the behavior of the considered networks, which learned the composition of rotation and scaling.

$$\Psi 2(z) = az \tag{5.3}$$

where $a = \alpha e^{i\tau}$ in Eq. (5.3) rotates the vector z by τ in counterclockwise direction and dilates or contracts it by a factor α.

This example explores the behavior of different network and learning algorithm for mapping $\Psi 2$. All networks are run up to 6,500 epochs with $C - BP$ ($\eta = 0.003$) and 1,000 epochs with $\boldsymbol{C}$RPROP algorithm ($\mu^- = 0.4, \mu^+ = 1.2, \Delta_{\min} = 10^{(-6)}, \Delta_{\max} = 0.005, \Delta_0 = 0.01$). The 21 learning input patterns lying on a line $y = x - 0.1$, $(-0.9071 \leq x \leq 0.507)$ are contracted by $\alpha = 1/2$ and rotated over $\pi/2$ radians anticlockwise to output patterns $y = -x - 0.5$, $(-0.553 \leq x \leq 0.153)$, at reference point $(-0.2, -0.3)$. The input test points lying on the ellipse $\frac{(x+0.2)^2}{a^2} + \frac{(y+0.3)^2}{b^2} = 1$ would hopefully be mapped to points lying on $\frac{(x+0.2)^2}{(b/2)^2} + \frac{(y+0.3)^2}{(a/2)^2} = 1$ at reference $(-0.2, -0.3)$, where $a = 0.7, b = 0.3$. Transformations displayed in Fig. 5.3 show the generalization over ellipse.

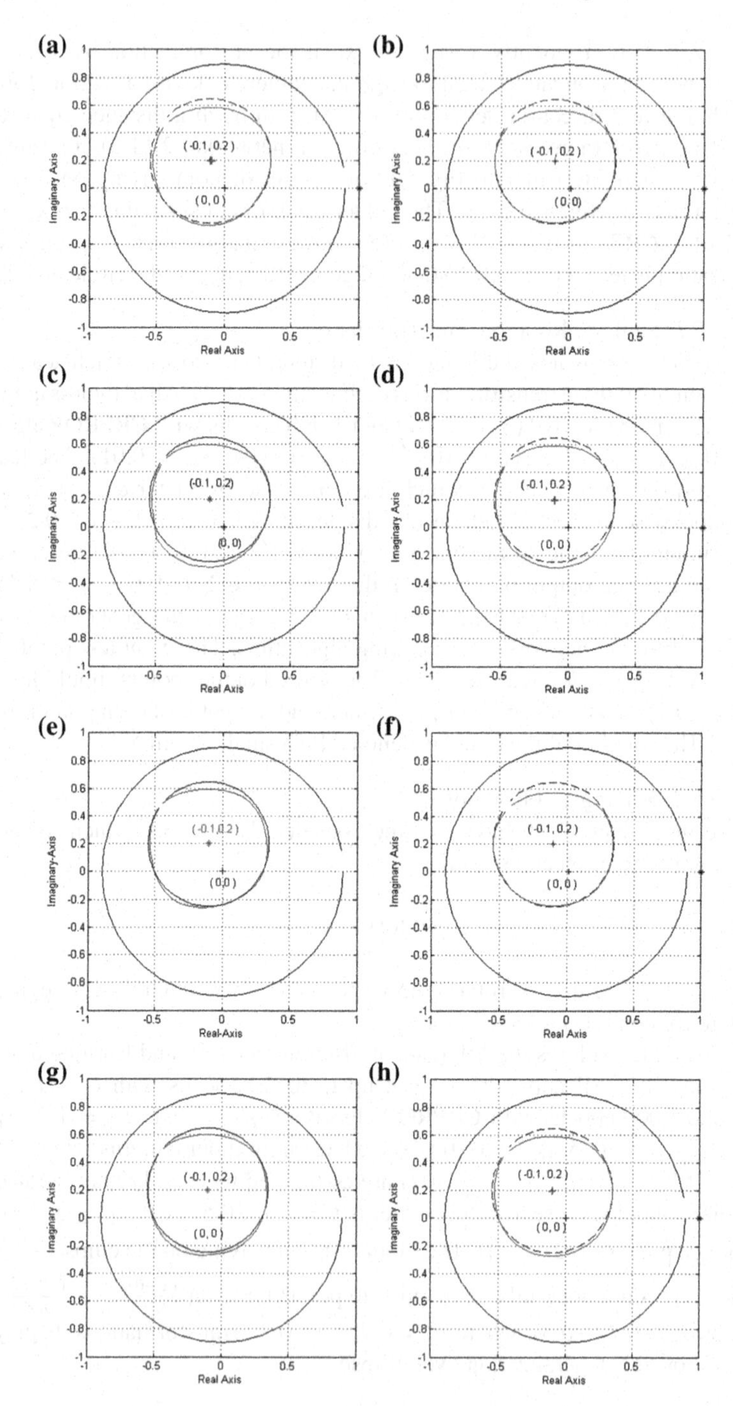

Fig. 5.2 Scaling, rotation, and translation with **a** *C*MLP, *C*BP, **b** *C*MLP, *C*RPROP, **c** *C*RPN, *C*BP, **d** *C*RPN, *C*RPROP, **e** *C*RSS, *C*BP, **f** *C*RSS, *C*RPROP, **g** *C*RSP, *C*BP, **h** *C*RSP, *C*RPROP

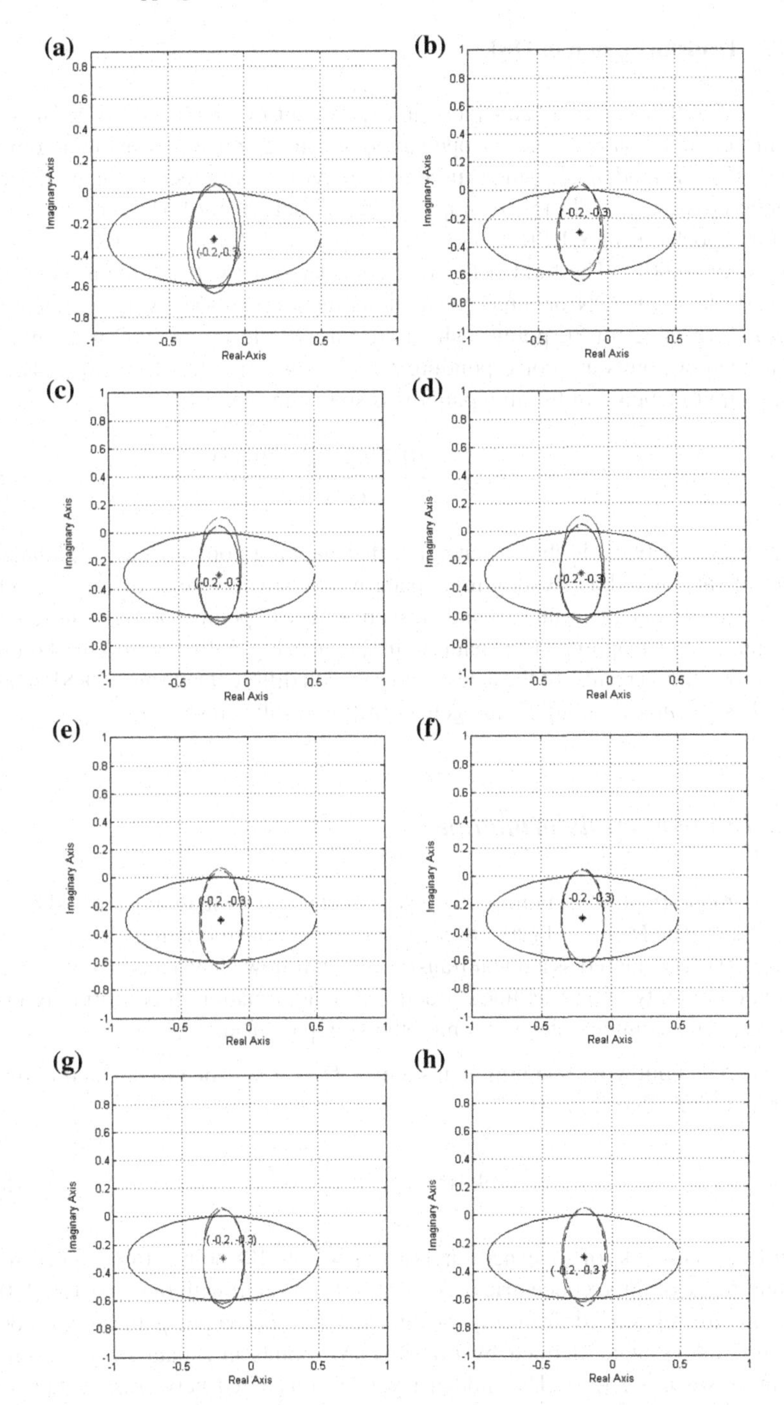

Fig. 5.3 Scaling and rotation with **a** *C*MLP, *C*BP, **b** *C*MLP, *C*RPROP, **c** *C*RPN, *C*BP, **d** *C*RPN, *C*RPROP, **e** *C*RSS, *C*BP, **f** *C*RSS, *C*RPROP, **g** *C*RSP, *C*BP, **h** *C*RSP, *C*RPROP

5.2.1.1 Transformation of Disk

Example 5.3 This example considers an experiment in which a set of points on a disk in a complex plane ‘z’ get mapped to points on an image disk in other complex plane ‘zz’. This kind of mapping is used to study viscous flow across the bodies with different cross section. In this example, different networks (2-M-2) are trained up to 2,000 epochs using ***C***RPROP ($\mu^- = 0.5$, $\mu^+ = 1.2$, $\Delta_{\min} = 10^{(-6)}$, $\Delta_{\max} = 0.04$, $\Delta_0 = 0.001$). The hidden layer of considered networks contain one ***C***RSP or ***C***RSS or two ***C***RPN or three conventional neurons, respectively. The learning patterns form a set of 36 points, which are on a circle $|z| = 0.62$ referenced at origin. Output points are corresponding value of $\Psi 3(z)$ defined by Eq. (5.4). The input–output mapping of training patterns is shown in Fig. 5.4a.

$$zz = \Psi 3(z) = \frac{(0.2\,z + 0.2 + 0.3j)}{(j + 0.4)} \tag{5.4}$$

The trained network is able to generalize this mapping from z-plane to zz-plane for circles of varying radius. Testing input patterns contain six circles of varying radius from 0.2 to 0.9 at regular interval and each circle has 72 points on its circumference. The input–output mapping of test patterns in shown in Fig. 5.4b. Figure 5.4c–f present the transformation results of these test patterns with different neurons. ***C***RSP neuron shows best accuracy among all the generalization results presented.

5.2.2 Bilinear Transformation

Bilinear transformation is an important class of elementary mapping studied by Augustus Ferdinand Mobius (1790–1868). The Bilinear transformation (*Mobius Transformation*) is considered as a linear transformation followed by a reciprocal transformation, commonly known as linear fractional transformation. It conformally maps one-one from a complex plane ‘z’ onto other complex plane ‘zz’.

Example 5.4 Following mapping maps a disk D_z: $|z| < 1$ one to one and onto the upper half plane $Im(zz) > 0$.

$$zz = \Psi 4(z) = j \times \frac{(1 - z)}{(1 + z)} \tag{5.5}$$

Different networks of novel neurons are trained with 108 points (First input) on the circumference of three concentric disk (36 points on each) with radius 0.1, 0.3, 0.5. All disks are referenced at origin (Second input). The output patterns are corresponding values of zz defined by Eq. (5.5). All points in a data set are normalized in between −1 to 1. The hidden layer of considered networks contain four ***C***RSP or ***C***RSS, or eight ***C***RPN or conventional neurons, respectively. Normalized input–output mapping for training patterns is shown in Fig. 5.5a. Learning

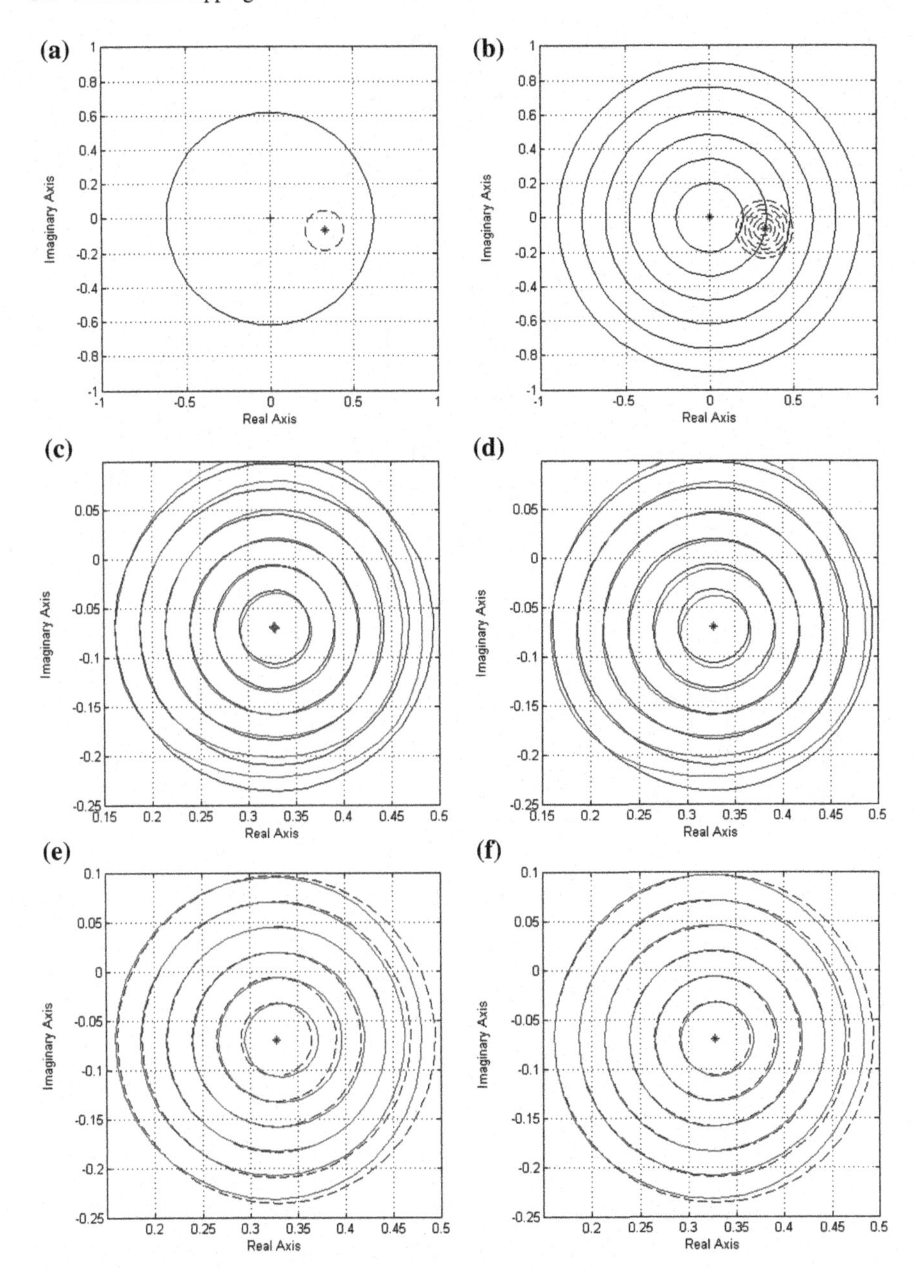

Fig. 5.4 **a** Training input–output patterns. **b** Test input–output patterns. Transformation of test patterns with **c** *C*MLP, *C*RPROP, **d** *C*RPN, *C*RPROP, **e** *C*RSS, *C*RPROP, **f** *C*RSP, *C*RPROP

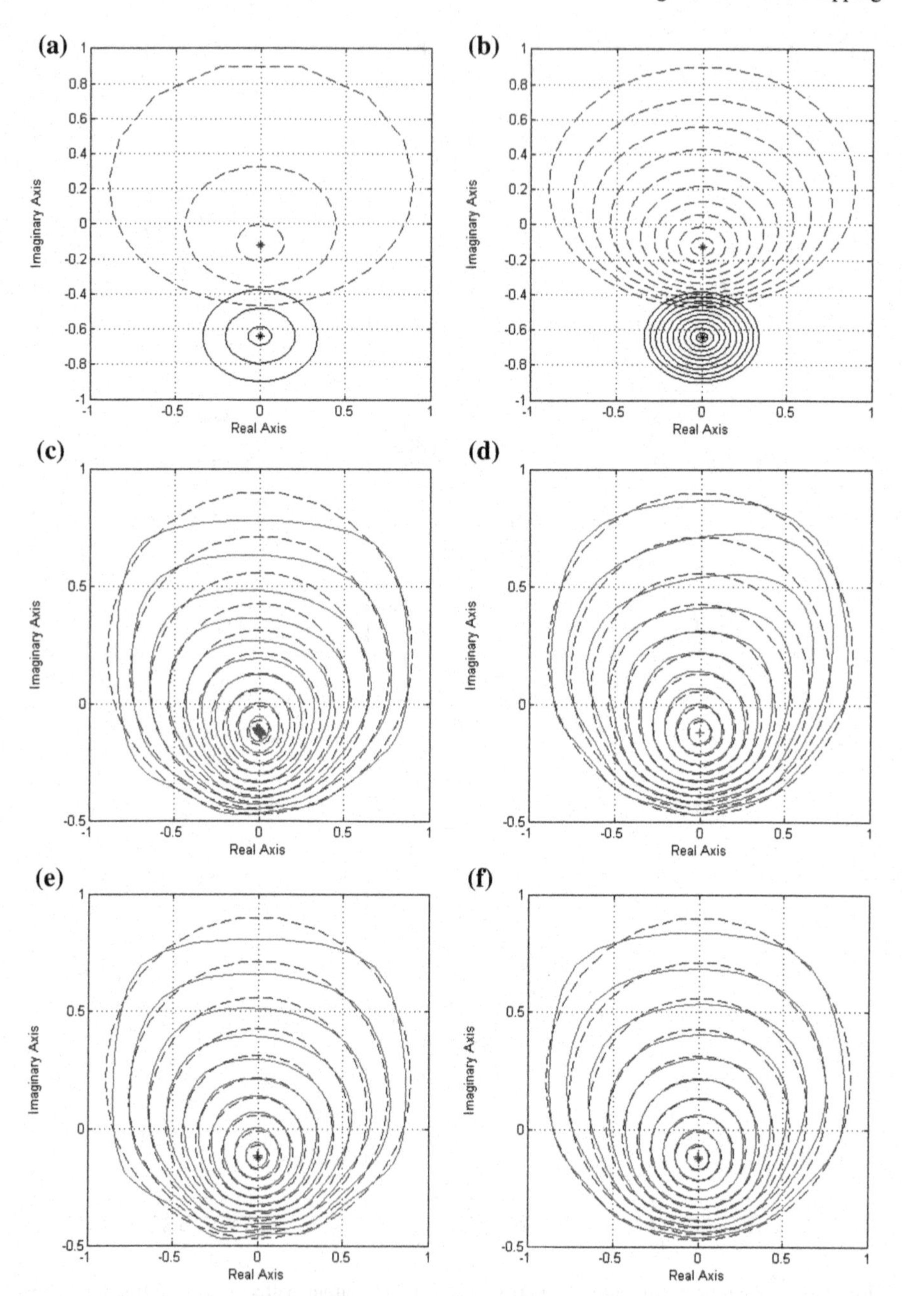

Fig. 5.5 **a** Normalized input–output mapping for training. **b** Normalized input–output mapping for testing. Bilinear Transformation of test patterns with **c** ***C***MLP, ***C***RPROP, **d** ***C***RPN, ***C***RPROP, **e** ***C***RSS, ***C***RPROP, **f** ***C***RSP, ***C***RPROP

in different networks was done up to 12,000 epochs using $\mathbb{C}$RPROP algorithm ($\mu^{-} = 0.5, \mu^{+} = 1.2, \Delta_{\min} = 10^{(-6)}, \Delta_{\max} = 0.001, \Delta_0 = 0.0005$). The $\mathbb{C}$BP failed to train any network for bilinear transformation, as saturation is observed in learning. However, with $\mathbb{C}$RPROP algorithm, training, and generalization has been successfully achieved. The trained networks are able to generalize this mapping from 'z' plane to 'zz' plane over varying values of radius of disks from 0.05 to 0.5 at 10 regular intervals. Normalized test input–output mapping in shown in Fig. 5.5b. Figure 5.5c–f present the transformation results of these test patterns with different networks. The output of $\mathbb{C}$RSP neuron based network displayed in Fig. 5.5f shows that it gives best generalization as compared to other networks. The superiority of $\mathbb{C}$RSP neuron is again seen in this case also.

5.3 Concluding Remarks

The linear and bilinear transformations are frequently employed and the example mentioned here is but a small application of the innumerable existent. As these functions are most frequently applied, the CVNN along with conventional and higher order neurons are used to address them to study the convergence. The mapping problems demonstrated in this chapter represent typical functions encountered in practice. Results in mapping applications demonstrate that the standard CVNN can be replaced by the higher order neurons presented in earlier chapter for better performance. The complex mapping was found to be sensitive to the normalization of the input and output patterns. The $\mathbb{C}$AF demand that the data to be mapped be restricted to a range or otherwise, the functions' effective contribution to the weights diminishes for larger values of the parameter (as the functions become more flat for large arguments and hence less slope). This restriction for implementing the CVNN algorithms with the bilinear transformation manifests in the form of a constrained range on the parameters. For circles of a large radius and lying beyond the unit circle, a normalization factor must be introduced to restrict the whole output to within the unit circle before using the CVNN with this activation function.

Simulations on these complex-valued problems clearly presents the robustness and functional superiority of proposed neurons over conventional neurons. These neurons make it possible to solve the problems using a smaller network and fewer learning parameters. Besides, they have also demonstrated faster learning and better approximation. This has been found in few problems that ANN units are trapped at early saturation with $\mathbb{C}$BP algorithm, which preclude any significant improvement in the training weights. This causes an unnecessary increase in the number of iterations required to train an ANN. This situation is serious in bilinear transformation, where $\mathbb{C}$BP failed to converge. We have obtained a fine result for this problem with $\mathbb{C}$RSP and $\mathbb{C}$RPROP. Moreover, the modified $\mathbb{C}$RPROP has thoroughly demonstrated better performance with drastic reduction in training epochs. Efficient solution provided in conformal mapping ensures the ability of complex domain neurons to process magnitude and phase of data/signals properly.

References

1. Brown, J.W., Churchill, R.V.: Complex Variables and Applications, 7th edn. McGraw-Hill, New York (2003)
2. Nitta, T.: An analysis of the fundamental structure of complex-valued neurons. Neural Process. Lett. **12**, 239–246 (2000)
3. Tripathi, B.K., Kalra, P.K.: The novel aggregation function based neuron models in complex domain. Soft Comput. **14**(10), 1069–1081 (2010)
4. Saff, E.B., Snider, A.D.: Fundamentals of Complex Analysis with Applications to Engineering and Science. Prentice-Hall. Englewood Cliffs (2003)
5. Nitta, T.: An extension of the back-propagation algorithm to complex numbers. Neural Netw. **10**(8), 1391–1415 (1997)

Chapter 6
Neurocomputing in Space

Abstract The high-dimensional neural network is becoming very popular in almost every intelligence system design, just to name few, computer vision, robotics, biometric identification, control, communication system, and forecasting are the scientific and engineering fields that take advantage of artificial neural networks (ANN) to emulate intelligent behavior. In computer vision the interpretation of 3D motion, 3D transformations, and 3D face or object recognition are important tasks. There have been many methodologies to solve them, but these methods are time consuming and weak to noise. The advantage of using neural networks for object recognition is the feasibility of a training system to capture the complex class conditional density of patterns. It will be desirable to explore the capabilities of ANN that can directly process three-dimensional information. This article discusses the machine learning from the view points of 3D vector-valued neural network and corresponding applications. The learning and generalization capacity of high-dimensional ANN is confirmed through diverse simulation examples.

6.1 3D Vector-Valued Neuron

The advantages with a neural network include robustness, ability to learn, generalize, and separate complicated classes [1, 2]. In 3D vector-valued neural network, the input–output signals and threshold are 3D real-valued vectors, while weights associated with connections are 3D orthogonal matrices. The weights are assumed to be orthogonal matrices because this assumption is a natural extension of the weights of the complex-valued neuron. We will present few illustrative examples to show how a 3D vector-valued neuron can be used to learn 3D motion and used in 3D face recognition. The proposed 3D motion interpretation system is trained using only few set of points lying on a line in the 3D space. The trained system is capable of interpreting 3D motion consisting of several motion components over unknown 3D objects. Face recognition is the preferred mode of identity authentication [3–5]. The facial features have several advantages over other six biometric attributes considered by Hietmeyer [6]. It is natural, robust, and uninstructive. It cannot be forgotten or mislaid like other document of identification. Most of the face recognition techniques

B.K. Tripathi, *High Dimensional Neurocomputing*,
Studies in Computational Intelligence 571, DOI 10.1007/978-81-322-2074-9_6

have used 2D images of human faces. However, 2D face recognition techniques are known to suffer from the inherent problems of illumination and structural variation, and are sensitive to factors such as background, change in human expression, pose, and aging [7]. Utilizing 3D face information was shown to improve face recognition performance, especially with respect to these variations [8, 9].

The complexity of ANN depends on the number of neurons and learning algorithm. The higher the complexity of an ANN is, the more computations and memory intensive it can be. The number of neurons to be used in an ANN is a function of the mapping or the classifying power of the neuron itself [2, 10]. Therefore, in case of high- dimensional problem, it is imperative to look for higher-dimensional neuron model that can directly process the high-dimensional information. It will serve as a building block for a powerful ANN with fewer neurons. Various researchers have independently proposed extension of real-valued neuron (one dimension) to higher dimension [2, 11]. Most of them have followed natural extension of number field like real number (one dimension), complex number (two dimension), 3D real-valued vectors (three dimension), quaternion (four dimension), etc., for representation of higher-dimension neurons. Therefore, it will be worthwhile to explore the capabilities of the 3D vector-valued neurons in function mapping and pattern classification problems in 3D space. The activation function for 3D vector-valued neuron can be defined as 3D extension of real activation function. Let $V = [V_x, V_y, V_z]^T$ be the net internal potential of a neuron then its output is defined as:

$$Y = f(V) = [f(V_x), f(V_y), f(V_z)]^T \tag{6.1}$$

6.1.1 Learning Rule

In our multilayer network, we have considered three layers, first is of inputs, second layer is only hidden layer, and an output layer. A three-layer network can approximate any continuous nonlinear mapping. In 3D vector-valued neural network, the bias values and input–output signals are all 3D real-valued vectors, while weights are 3D orthogonal matrices. All the operations in such a neural network are scaler matrix operations. A 3D vector-valued back-propagation algorithm (3DV-BP) is considered here for training a multilayer network, which is natural extension of complex-valued back-propagation algorithm [10, 12]. It has ability to learn 3D motion as complex-BP can learn 2D motion [13].

In a three-layer network (L-M-N), the first layer has L inputs (I_l), where $l = 1..L$, the second and the output layer consist M and N vector-valued neurons, respectively. By convention, w_{lm} is the weight that connects lth neuron to mth neuron and $\alpha_m = [\alpha_{mx}, \alpha_{my}, \alpha_{mz}]$ is the bias weight of mth neuron. $\eta \in [0, 1]$ is the learning rate and f' is derivative of a non-linear function f. Let V be net internal potential and Y be the output of a neuron. Let e_n be the difference between actual and desired value at nth output, where $|e_n| = \sqrt{e_n^{x2} + e_n^{y2} + e_n^{z2}}$ and $e_n = [e_n^x, e_n^y, e_n^z]^T = Y_n - Y_n^D$.

$$W_{lm} = \begin{vmatrix} w_{lm}^x & -w_{lm}^y & 0 \\ w_{lm}^y & -w_{lm}^x & 0 \\ 0 & 0 & w_{lm}^z \end{vmatrix} \qquad W_{mn} = \begin{vmatrix} w_{mn}^x & 0 & 0 \\ 0 & w_{mn}^y & -w_{mn}^z \\ 0 & w_{mn}^z & w_{mn}^y \end{vmatrix}$$

where $W_{lm}^z = \sqrt{(w_{lm}^x)^2 + (w_{lm}^y)^2}$ and $W_{mn}^x = \sqrt{(w_{mn}^y)^2 + (w_{mn}^z)^2}$

The net potential of mth neuron in hidden layer can be given as follows:

$$V_m = \sum_l w_{lm} I_l + \alpha_m \tag{6.2}$$

The activation function for 3D vector-valued neuron is 3D extension of real activation function and defined as follows:

$$Y_m = f(V_m) = \left[f(V_m^x), f(V_m^y), f(V_m^z)\right]^T. \tag{6.3}$$

Similarly,

$$V_n = \sum_m w_{mn} Y_n + \alpha_n \quad \text{and} \quad Y_n = f(V_n) = \left[f(V_n^x), f(V_n^y), f(V_n^z)\right]^T. \tag{6.4}$$

$$Y_n = f(V_n) = \left[f(V_n^x), f(V_n^y), f(V_n^z)\right]^T. \tag{6.5}$$

The mean square error function can be defined as:

$$E = \frac{1}{N} \sum_n |e_n|^2 \tag{6.6}$$

In 3D vector version of back-propagation algorithm (3DV-BP) the weight update equation for any weight is obtained by gradient descent on error function:

$$\Delta w = \eta \begin{vmatrix} -\frac{\partial E}{\partial w^x} & -\frac{\partial E}{\partial w^y} & -\frac{\partial E}{\partial w^z} \end{vmatrix}^T$$

then, weights and bias in output layer can be updated as follows:

$$\Delta \alpha_n = \eta \begin{vmatrix} e_n^x \cdot f'(V_n^x) \\ e_n^y \cdot f'(V_n^y) \\ e_n^z \cdot f'(V_n^z) \end{vmatrix}$$

$$\begin{vmatrix} \Delta w_{mn}^y \\ \Delta w_{mn}^z \end{vmatrix} = \eta \begin{vmatrix} \frac{w_{mn}^y}{w_{mn}^x} Y_m^x & Y_m^y & Y_m^z \\ \frac{w_{mn}^z}{w_{mn}^x} Y_m^x & -Y_m^z & Y_m^y \end{vmatrix} \begin{vmatrix} e_n^x \cdot f'(V_n^x) \\ e_n^y \cdot f'(V_n^y) \\ e_n^z \cdot f'(V_n^z) \end{vmatrix}$$

Similarly, weights and bias in hidden layer neuron can be updated as follows

$$\begin{vmatrix} \Delta\alpha_m^x \\ \Delta\alpha_m^y \\ \Delta\alpha_m^z \end{vmatrix} = \frac{\eta}{N} \begin{vmatrix} f'(V_m^x) & 0 & 0 \\ 0 & f'(V_m^y) & 0 \\ 0 & 0 & f'(V_m^z) \end{vmatrix} \sum_n \begin{vmatrix} w_{mn}^x & 0 & 0 \\ 0 & w_{mn}^y & w_{mn}^z \\ 0 & -w_{mn}^z & w_{mn}^y \end{vmatrix} \begin{vmatrix} e_n^x \cdot f'(V_n^x) \\ e_n^y \cdot f'(V_n^y) \\ e_n^z \cdot f'(V_n^z) \end{vmatrix}$$

$$\begin{vmatrix} \Delta w_{lm}^x \\ \Delta w_{lm}^y \end{vmatrix} = \begin{vmatrix} I_l^x & I_l^y & \frac{w_{lm}^x}{w_{lm}^z} I_l^z \\ -I_l^y & I_l^x & \frac{w_{lm}^y}{w_{lm}^z} I_l^z \end{vmatrix} \begin{vmatrix} \Delta\alpha_m^x \\ \Delta\alpha_m^y \\ \Delta\alpha_m^z \end{vmatrix}$$

6.2 Learning 3D Motion

This section presents different 3D motions of objects in the physical world. The motion in a 3D space may consists of a 3D *scaling, rotation, and translation* and composition of these three operations. These three basic class of transformations convey dominant geometric characteristic of mapping. Such a mapping in space preserves the angles between oriented curves, and the phase of each point on the curve is also maintained during motion of points. As described in [1, 14], the complex-valued neural network enables to learn 2D motion of signals, hence generalizes conformal mapping on plane. Similarly, a 3D vector-valued neural network enables to learn 3D motion of signals, and will provide generalization of mappings in space. In contrast, a neural network in a real domain administers 1-D motion of signals, hence does not preserve the amplitude as well as phase in mapping. This is the main reason as to why a high dimensional ANN can learn high dimensional mapping, while equivalent real-valued neural network cannot [11].

In order to validate the proposed motion interpretation system, various simulations are carried out for learning and generalization of high dimensional mapping. It has the capacity to learn 3D motion patterns using set of points lying on a line in the 3D space and generalize them for motion of an unknown object in the space. We have used a 2-6-2 structure of 3D vector-valued neural network in all experiments of this section, which transform every input point (x, y, z) into another point (x', y', z') in the 3D space. First input of input layer takes a set of points lying on the surface of a object and second input is the reference point of input object. Similarly, the first neuron of output layer gives the surface of transformed object and second output is its reference point. Empirically, it is observed that considering reference point yields better testing results. The input–output values are with in the range $-1 \leq x, y, z \leq 1$. In all simulations, the training input–output patterns are the set of points lying on a straight line with in the space of unit sphere ($0 \leq$ radius ≤ 1), centered at the origin and all the angles vary from 0 to 2π. Figure 6.1a presents an example input–output mapping of training patterns. The following few examples depict the generalization ability of such trained network over standard geometric objects.

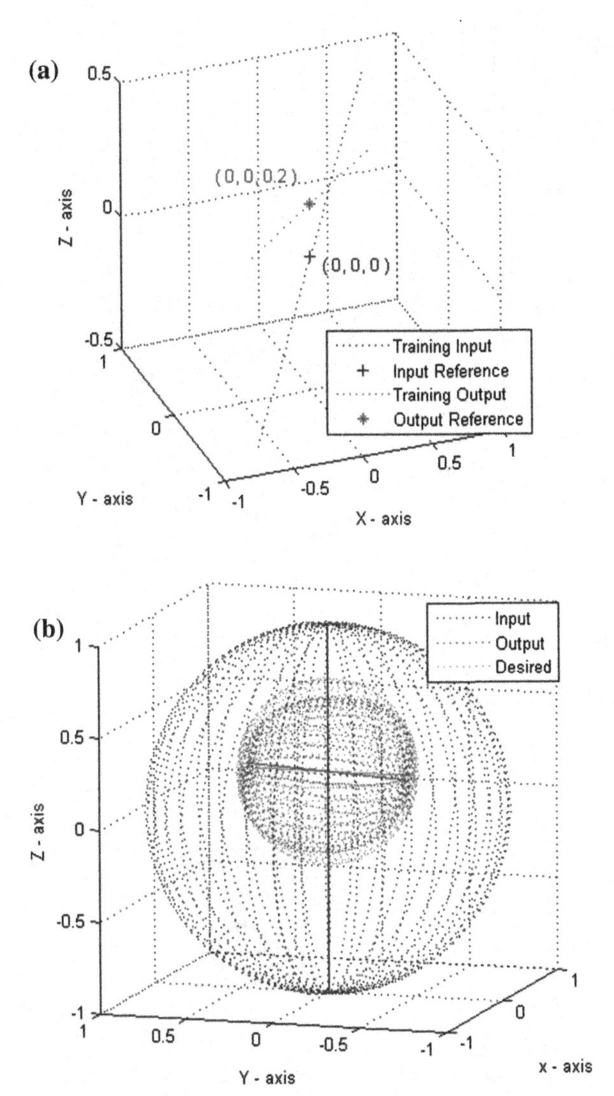

Fig. 6.1 **a** *Training patterns* mapping shows scaling by factor $1/2$, angle of rotation $\pi/2$ and displacement by (0, 0, 0.2); **b** The testing over a 3D object

Example 6.1 A neural network based on 3D vector-valued neurons has been trained for the composition of all three transformation. Training input–output patterns are shown in Fig. 6.1a. The generalization ability of such a trained network is tested over sphere (1681 data pints). Figure 6.1b presents the generalization ability of trained network. All patterns in output are contracted by factor $1/2$, rotated over $\pi/2$ radians clockwise and displaced by (0, 0, 0.2). This example demonstrate the generalization ability of considered network in interpretation of object motion in 3D space.

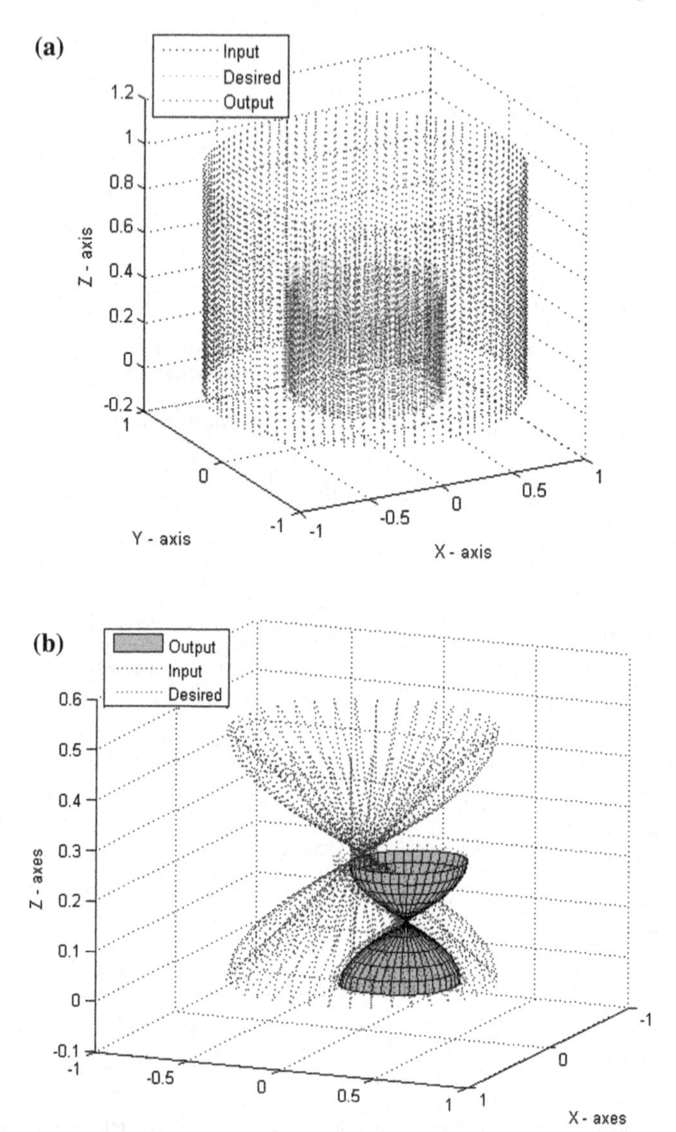

Fig. 6.2 **a** Similarity transformation in 3-D; **b** Scaling and translation in 3-D space

Example 6.2 In this experiment the network is trained for input–output mapping over a straight line for similarity transformation (scaling factor $1/2$) only. The generalization ability of such trained network is tested over cylinder containing 202 data points. The transformation result in Fig. 6.2a shows the excellent generalization with proposed methodology.

Example 6.3 A neural network based on 3D vector-valued neurons has been trained with line for the composition of scaling and translation. Figure 6.2b presents the generalization ability of this trained network. There are 451 test data points on cylinder. All points in 3D are contracted by factor $1/2$ and displaced by $(0, 0, 0.2)$. Results bring out the fact that given neural network is able to learn and generalize 3D motion.

6.3 Point Clouds of Objects in Practical Application

The CVNN was applied to the various benchmarks and mapping problems in the previous chapters. In the present chapter, the problem of 3D object classification is performed through the sorting of point clouds of objects using vector-valued neural network. The wide spectrum of problems in science and engineering deal in constructing and analyzing surfaces from clouds of points. A typical problem in the area has a data set at hand obtained by running a scanning-equipment across the object of interest. The scan must be performed in an orderly fashion to ensure the data points are well organized (and do not appear at irregular intervals or appear disorderly). After the operation, a point cloud of the object gets generated that is subject to further analysis. So the algebraic functions (polynomials) may also be employed in many cases to fit point clouds. It must be acknowledged, however, that the actual form of the surface of object if known would give a more accurate picture of the surface and the following analysis more correctly placed. The problem of object classification is a problem of mapping, which refers to sorting a set of object into categories with predefined characteristics, in which the number of classes is usually fixed. To start with, the 3D vector-valued back-propagation algorithm (3DV-BP) may be run to train the neural network architecture of appropriate size. To keep the study uniform, a same size of neural network architecture is chosen with uniform learning rate and the training epochs. To test the neural network another sets of object's surfaces are constructed.

6.4 3D Face Recognition

Biometrics can be defined as the automated use of physiological (face, finger prints, periocular, iris, Oculomotor plant characteristic, and DNA) and behavioral (signature and typing rhythms) characteristics for verifying the identity of living person. The physiological features are often non-alterable except severe injury, while behavioral features may fluctuate due to stress, fatigue, or illness. Face recognition is one of the few biometric methods that possess the merits of both high accuracy and low intrusiveness. It is also one of the most acceptable biometrics because a human face is always bare and often used in their visual interactions. It is a potential identify of a person without document for identification.

In early methods for 3D face recognition curvatures and surface features, kept in a cylindrical coordinate system, were used. Moreno et al. found that curvature and line features perform better than area features [15]. Point cloud is the most primitive 3D representation for faces and Housdroff distance has been used for matching the point clouds in [16]. The base mesh is also used for alignment in [17], where features are extracted from around landmark points and nearest neighbor, and after that PCA is used for recognition. In [8] the analysis-by-synthesis approach that uses morphable model is detailed. The idea is to synthesize a pose and illumination corrected image pair for recognition. Depth maps have been used in 3D imaging applications [18]. The depth map construction consists of selecting a view point and smoothing the sampled depth values. Most of the work that uses 3D face data use a combination of representations. The enriched variety of features, when combined with classifiers with different statistical properties, produce more accurate and robust performance. As a result of fast development in 3D imaging technology, there is strong need to address them using high-dimensional neural networks.

6.4.1 Normalization

3D linear transformation has been considered for normalization of 3D faces. Its purpose is to align each face on a same scale and at same orientation. In order to make the standard alignment for facial features, the origin is translated to the nose tip. It is assumed that the scanned 3D face data are of front part of face and almost straight (variation 40°–50° allowed) and accordingly it is translated. Logically, nose tip is the peak of a face, and hence can have maximum Z-coordinate value. Therefore, the Z coordinate ($Z_{\max}$) on 3D face data is searched, and their corresponding X, Y coordinates, say $Z_{\max} \cdot X$ and $Z_{\max} \cdot Y$ are determined. Now, origin is translated to nose tip as follows:

$$\hat{X} = X - Z_{\max} \cdot X \qquad \hat{Y} = Y - Z_{\max} \cdot Y \qquad \hat{Z} = Z - Z_{\max}$$

Thus, we obtain a face data with the origin on the nose tip. Now face is rotated about Y and X axis. For Y axis rotation, replacement in coordinates can be done as follows:

$$\begin{Bmatrix} z' = z \times cos\theta - x \times sin\theta \\ x' = z \times sin\theta + x \times cos\theta \\ y' = \ \ y \end{Bmatrix} \qquad R_y(\theta) = \begin{vmatrix} cos\theta & 0 & sin\theta & 0 \\ 0 & 1 & 0 & 0 \\ sin\theta & 0 & cos\theta & 0 \\ 0 & 0 & 0 & 1 \end{vmatrix}$$

where, $\theta = tan^1(x'/z')$ is an angle to which vector is rotated about Y axis. Similarly, the face is rotated about the X axis. Once the nose tip is identified, one can search in the y direction to determine the nose dip. Both nose tip and nose dip must lie on the same line. The scaling of all the faces has been done by taking distance between the nose tip and nose dip along y-axis and nose tip as the origin of the coordinate system.

6.4.2 Recognition

Recent development in computer technology and call for better security applications have brought biometrics into focus. The signature, handwriting, and fingerprint have a long history. More recently voice, retinal scan, iris scan, and face information are considered for biometrics. When deploying a biometrics-based system, we consider its accuracy, cost, ease of use, whether it allows integration with other systems, and the ethical consequences of its use. This chapter focus on 3D pattern classification using neural network. Our method has successfully performed recognition irrespective of variability in head pose, direction, and facial expressions. We present here two illustrative examples to show how a neural network based on 3D real-valued neurons can be used to learn and recognize point cloud data of 3D faces. A 1-2-1 network of vector-valued neurons was used in following two experiments. The proposed pattern classifier structure involves the estimation of learning parameters (weights) which are stored for future testing. It is more compact and can be easily communicated to humans than learned rules.

Example 6.4 The 3D vector-valued neural network was trained by a face (Fig. 6.3a) from first set of face data (Fig. 6.3). This face data contains five faces of different persons, where each face contains 6397 data pints. Table 6.1 presents the testing error yielded by trained network for all five faces (Fig. 6.3). The testing error for four other faces is much higher in comparison to the face that is used in training. Thus, trained network recognize the face, which is taken in training, and reject four faces of other persons. Results bring out the fact that this methodology is able to learn and classify the 3D faces correctly.

Example 6.5 In this example, the considered network was trained by first face (Fig. 6.4a) from the face set (Fig. 6.4). This face set contains five faces of same person with different orientation and poses. Each face contains 4663 data pints. Table 6.2 presents the testing error yielded by trained network for all five faces (Fig. 6.4). The testing error for four other faces is also minimum and comparable to the face, which is used for training. Thus, trained network recognize all faces of same person. Thus, considered methodology has successfully performed recognition irrespective of variability in head pose and orientation.

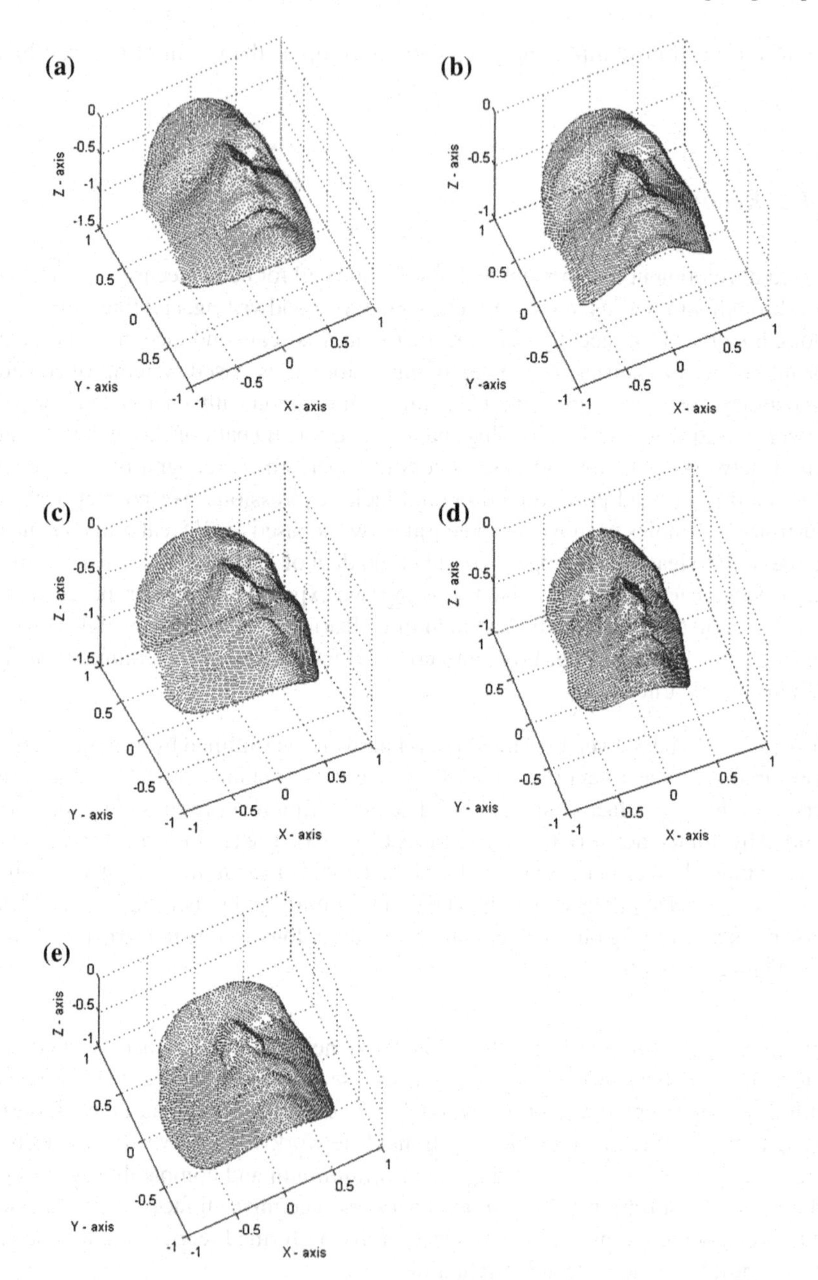

Fig. 6.3 Five faces different persons considered in Example 6.4

Table 6.1 Comparison of testing error of face set (Fig. 6.3)

Test face	6.3a	6.3b	6.3c	6.3d	6.3e
Test error	7.79e-05	9.34e-01	2.02e-00	5.73e-02	2.61e-01

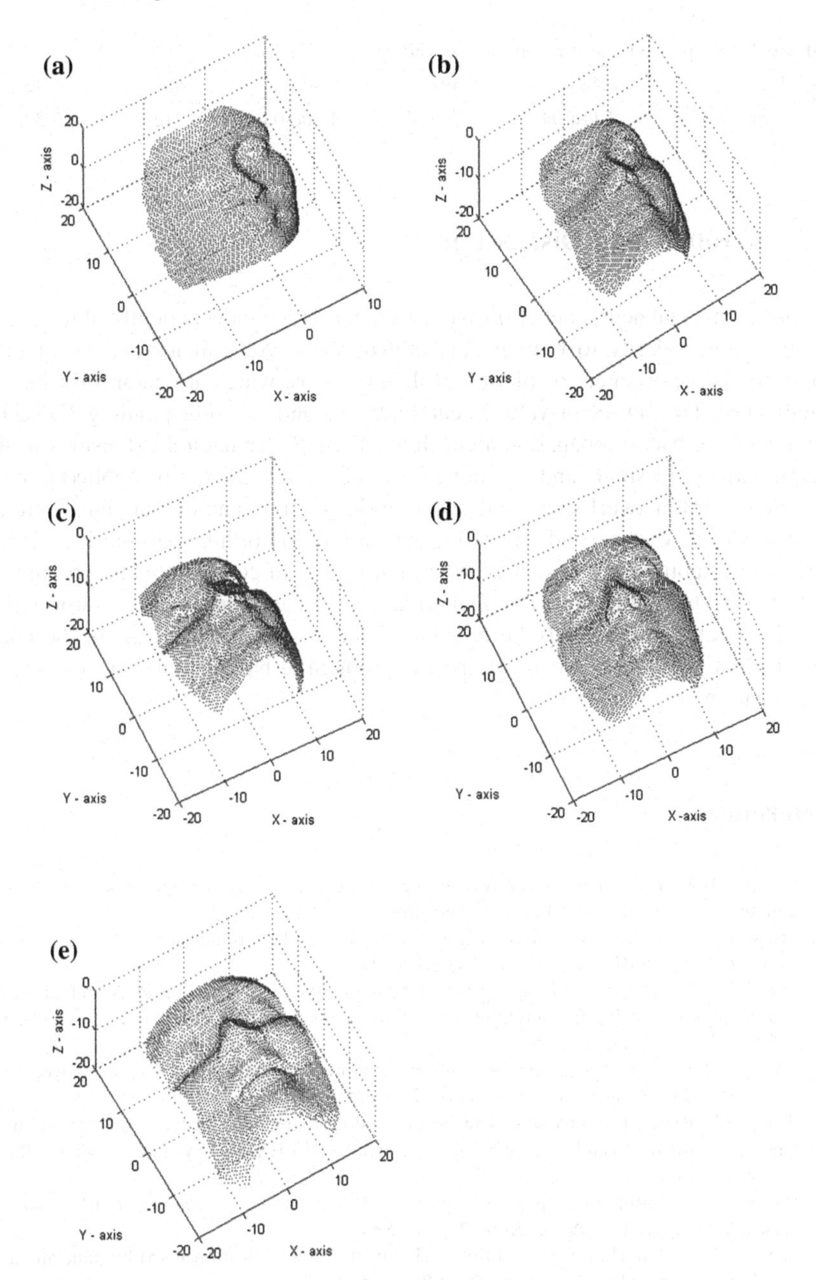

Fig. 6.4 Five faces of same person with different orientation and poses, Example 6.5

Table 6.2 Comparison of testing error of second face set (Fig. 6.4)

Test face	6.4a	6.4b	6.4c	6.4d	6.4e
Test error (MSE)	3.11e-04	6.10e-03	4.90e-03	6.21e-04	7.32e-04

6.5 Inferences and Discussion

High-dimensional neural networks have been developed independently of their similarity or unsimilarity to biological neural network. Nevertheless, many of their applications are oriented to solution of the problems, which are natural for human intelligence. The 3D vector-valued neural network and the corresponding 3D vector version of the back-propagation algorithm (3DV-BP) are natural extensions of the neural network of single and two dimensions. The 3DV-BP can be applied to multilayered neural networks whose threshold values, input, and output signals are all 3D real valued vectors, and whose weights are all 3D orthogonal matrices. It has been well established in this chapter through variety of computational experiments that 3DV-BP has ability to learn 3D motion. It is hence clear that phase information of each point can effectively be maintained along with the decrease in the number of input parameters when solving the problem at hand—function mapping or classification.

References

1. Tripathi, B.K., Kalra, P.K.: The novel aggregation function based neuron models in complex domain. Soft Comput. **14**(10), 1069–1081. Springer, (2010)
2. Tripathi, B.K., Kalra, P.K.: On the learning machine for three dimensional mapping. Neural Comput. Appl. **20**(01), 105–111. Springer, (2011)
3. Oh, B.J.: Face recognition by using neural network classifiers based on PCA and LDA. In: Proceedings of the IEEE International Conference on Systems, Man and Cybernetics, pp. 1699–1703, 10–12 Oct 2005
4. Zhou, X., Bhanu, B.: Integrating face and gait for human recognition at a distance video. IEEE Trans. Syst. Man Cybern. **37**(5), 1119–1137 (2007)
5. Pantic, M., Patras, I.: Dynamics of facial expression: recognition of facial actions and their temporal segments from face profile image sequences. IEEE Trans. Syst. Man Cybern. **36**(2), 433–449 (2007)
6. Hietmeyer, R.: Biometric identification promises fast and secure processing of airline passengers. Int. Civil Aviat. Organ. J. **55**(9), 10–11 (2000)
7. O'Tolle, A.J., Abdi, H., Jiang, F., Phillips, P.J.: Fusing face-verification algorithms and humans. IEEE Trans. Syst. Man Cybern. **37**(5), 1149–1155 (2007)
8. Blanz, V., Vetter, T.: Face recognition based on fitting a 3D morphable model. IEEE Trans. PAMI **25**(9), 1063–1074 (2003)
9. Abate, A.F., Nappi, M., Riccio, D., Sabatino, G.: Letters 2D and 3D face recognition: a survey. Pattern Recognit. **28**, 1885–1906 (2007)
10. Tripathi, B.K., Kalra, P.K.: On efficient learning machine with root power mean neuron in complex domain. IEEE Trans. Neural Network **22**(05), 727–738 (2011). ISSN: 1045–9227

11. Nitta, T.: An analysis of the fundamental structure of complex-valued neurons. Neural Process. Lett. **12**, 239–246 (2000)
12. Piazza, F., Benvenuto, N.: On the complex backpropagation algorithm. IEEE Trans. Signal Process. **40**(4), 967–969 (1992)
13. Tripathi, B.K., Kalra, P.K.: Complex generalized-mean neuron model and its applications. Appl. Soft Comput. **11**(01), 768–777. Elsevier Science, (2011)
14. Nitta, T.: An extension of the back-propagation algorithm to complex numbers. Neural Networks **10**(8), 1391–1415 (1997)
15. Moreno, A.B., Sanchez, A., Velez, J.F., Daz, F.J.: Face recognition using 3D surface-extracted descriptors. In: Proceedings of the IMVIPC (2003)
16. Achermann, B., Bunke, H.: Classifying range images of human faces with Hausdorff distance. In: Proceedings of the ICPR, pp. 809–813 (2000)
17. Xu, C., Wang, Y., Tan, T., Quan, L.: Automatic 3D face recognition combining global geometric features with local shape variation information. In: Proceedings of the AFGR, pp. 308–313 (2004)
18. Lee, C.C., Chung, P.C., Tsai, J.R., Chang, C.I.: Robust radial basis function neural network. IEEE Trans. Syst. Man Cybern. B Cybern. 29(6), 674–685 (1999)

Chapter 7
Machine Recognition in Complex Domain

Abstract Machine recognition has drawn considerable interest and attention from researches in intelligent system design and computer vision communities over the recent past. Understandably there are a large number of commercial, law enforcement, control and forensic applications to this. We human beings have natural ability to recognize persons at a glance. Motivated by our remarkable ability, a series of attempts [1–4] have been made to simulate this ability in machines. The development of human recognition system in machines is quite difficult because the natural objects are complex, multidimensional, and corresponds to environmental changes [3, 5–7]. There are two important issues that need to be addressed in machine recognition: (1) how the features are adopted to represent an object under environmental changes and (2) how we classify an object image based on a chosen representation. Over the years, researches have developed a number of methods for feature extraction and classification. All of these, however, have their own merits and demerits. Most of the work is related to the real domain. The outperformance of complex-valued neuron over conventional neuron has been well established in previous chapters. Few researchers have recently tried multivariate statistical techniques in the complex domain, like complex principal component analysis (PCA) for 2D vector field analysis [8] and complex independent component analysis (ICA) for performing source separation on functional magnetic resonance imaging data [9, 10]. But, no attempts have been made to develop techniques for feature extraction using their concepts. This chapter presents formal procedures for feature extraction using unsupervised learning techniques in complex domain. Efficient learning and better precision in result offered by feature extractor and classifier, considering simulations in complex domain, figure out their technical benefits over conventional methods. Notably, the success of machine recognition is limited by variations in features resulting from the natural environment. These may be due to instrument distortion, acquisition in an outdoor environment, different noises, complex background, occlusion and illumination. A solid set of examples presented in this chapter demonstrate the superiority of feature representation and classification in complex domain.

B.K. Tripathi, *High Dimensional Neurocomputing*,
Studies in Computational Intelligence 571, DOI 10.1007/978-81-322-2074-9_7

7.1 State-of-Art: Machine Recognition

The machine recognition is one of the most distinguished facets jointly developed through the use of statistics and computational intelligence. Machine recognition is an interesting but challenging problem because it utilizes a mixture of techniques for feature representation and classification. It aroused interest of researchers of different background like psychology, information theory, ANN, and computer vision. The success of these techniques depends heavily on the particular choice of feature extractor and classifier. Feature representation or extraction involves the capturing of salient class-specific features in reduced dimension and redundancy elimination; as well providing enhanced discriminatory powers for future classification. The classifier or recognizer performs the recognition task. Human face recognition is the inherent capability of the human visual system. Though, humans do it effortlessly in spite of variations in facial features but it is not easy in computer vision terms. The facial features have several advantages over other six biometric attributes considered by Hietmeyer [11]. It is natural, passive, nonintrusive, and can be accessed comfortably. Therefore, the state-of-art discussions presented in this section; and further development and analysis of techniques in rest of the chapter have been done considering the face biometrics as a core for explanation.

Machine recognition results heavily depends on extracted features (a set of basis images) and classification methods. A standard machine recognition task should include three stages:

1. Preprocessing transforms the image data in a way where classification may prove convenient by exploitation of certain features. In this chapter, we include image resize, normalization or Hilbert transformation [12]. This ensures that all images are on equal platform.
2. Extraction of pertinent features to represent an object/image under environmental changes. They give new representation of objects based on the derived features.
3. Classification of objects/images based on new representation, into one of the classes.

7.1.1 Effective Feature Extraction

The literature in machine recognition is very vast and diverse [1, 2, 13]. A single system may utilize techniques motivated by different principles. This makes it difficult to categorize these systems, purely based on what techniques they use for feature extraction. The extraction of primitives of the face is very complex task. Variation among geometry, texture, and color of different areas of the images make it most discriminating. Motivated from psychological study of how humans use holistic and local features for recognition, two main approaches [14] for feature extraction have been evolved.

7.1.1.1 Structural or Model-Based Approach

It is based on extracting structural features that are local structure of images. It uses geometric relationships among these extracted structures. One of the popular methods of facial feature extraction is finding the geometry of face characteristics elements such as face contour, mouth, nose, eyes, etc. An explicit modeling of face variations has intrinsic physical relationship with real faces. Such model-based feature extraction (deformable templates, active contours, etc) is complicated and laborious. It is model fitting process and recognition results highly depend on the fitting results. The location and local statistics are fed into a structural classifier. The geometric features such as eyebrow thickness, nose vertical position and width, chin shape, etc, are extensively used for matching. Though, this approach is less sensitive to variations in illumination and to irrelevant information on an image; however, this approach of feature extraction is not much reliable. It has difficulties when appearances of features change significantly. For example, a smiling face and a frowning face are considered as two totally different image templates in rigid body sense. Similarly, faces with open eyes, close eyes and eyes with glasses. These techniques utilize pure geometry methods and require a lot of mathematics [14], hence computation intensive. One of the well-known methods in this category is the graph matching system [15, 16], which is based on dynamic link architecture. Another interesting approach is the hidden markov model (HMM) [17] and pseudo 2D HMM [18]. Blanz et al. [19] proposed a method based on 3D morphable model that encodes shape and texture in terms of model parameters. These parameters are used for face recognition.

7.1.1.2 Appearance Based or Holistic Approach

It is based on statistical approaches, where features are extracted from the whole image [6, 20–22]. The extracted feature are from global information, therefore they are affected from irrelevant information such as hair, shoulders or background which may affect the recognition results. Yet, the face may still be recognizable by viewing only the significant facial region [21]. In 1990, Kirby and Sirovich [23] have developed a technique for efficient representation of images using principal component analysis (PCA). Turk et al. [24] later used this technique for face recognition using eigenfaces and euclidean distance. Several successive method like eigenphases [25], appearance of local regions [26], Gabor-based PCA [27], linear or nonlinear discriminant analysis [28–30] use eigenspace transformation based on PCA. PCA transforms a number of possibly correlated variables into a smaller number of uncorrelated components. Recently, some applications of ICA [31–34] have been exploited in image processing and computer vision. It generalize the techniques of PCA, and proven to be more effective and superior to PCA in many applications. It defines the independent components from their linear mixture [35–37] of independent source signals. An important version of ICA, derived from the principle of optimal information transfer through sigmoid neuron has been used in [38].

Independent component representation captures the essential structure of data and make them more visible or accessible in many applications like feature extraction and signal separation.

Appearance-based feature extraction methods [14, 21] rely on techniques from statistical analysis to find the relevant characteristic of images. They transfer a recognition problem to image/signal space (principal or independent components) analysis problem, where well known methods like PCA [39], LDA [28], ICA [33, 34, 38] in real domain has been tried out. This requires sufficient representative data to sample the underlying distribution successfully. Feature extraction with PCA deals with second-order statistics only and removes correlation from the data (uncorrelates the data); while ICA accounts for higher-order statistics and aims at the removal of higher-order dependence. It is simple to find independent component from Gaussian data with PCA because for Gaussian data, the uncorrelated components are always independent. In real world, the data necessarily not following Gaussian distribution therefore uncorrelatedness in itself is not enough to separate the components, then ICA is preferred because ICA finds a representation in non-Gaussian distribution and yields the components that are statistically independent as possible. Thus, ICA provides more powerful data representation (basis vectors) than PCA [40, 41] and it has proved to be more effective than PCA for feature analysis [33, 38] in literature. Hence, it gives better results in partial occlusion or noisy environment.

This book focuses on the complex domain neurocompting, therefore it is highly desirable to develop techniques in complex domain in the role of feature extractor which may be compatible with tools of neurocompting in complex domain. It will be advantageous to develop a feature extraction technique using ICA in complex domain (*CICA*) and explore its performance over other statistical techniques. Complex PCA [8, 42] and complex ICA [9, 43, 44] have been introduced in literature to analyze the 2D vector fields and complex data. They have also been used to analyze the real data complexified first by Hilbert transformation [12, 42, 45]. This book develops feature extraction algorithms necessary for machine recognition using concepts of these statistical methods. This chapter is devoted in exploring the capability and effectiveness of 'feature space' computed with the feature selection schemes in complex domain. The new representation of images in the feature space may directly be used for classification with suitable classifier in complex domain.

7.1.2 Classifier Design

Classifiers play a significant role in the performance of recognition system apart from the feature extraction techniques. Statistical behavior of the feature vectors (new representation of patterns in lower dimensional subspace) is exploited to define decision regions corresponding to different classes. There are variety of discriminant functions (decision surface, distance metrices, separating hyperplane, threshold function, etc) available for classification. But in view of machine recognition, the discriminant functions are usually based on distance metrices. A nonlinear decision surface can also be formed using artificial neural networks. One can broadly observe the classifiers into two categories.

7.1.2.1 Distance Metric

Distance metric is widely used to measure the similarity between two features, when feature elements are extracted for different statistical properties of digital images. Some popular distance metrics described in the literature [34, 46] are L1 norm, L2 norm (Euclidean distance), cosine and Mahalanobis distance metric. The distance of the test image is compared with each of the training images already available. The test image is classified to belong to one of the classes which has least distance. Mean nearest neighbor and K-mean rule have also been explored. All these methods are computationally very demanding. The performance of different distance metrics also varies with source data distributions [7, 46]. It is difficult to find the best distance metric that fits the distribution of the underlying data.

7.1.2.2 Neural Network

Few neural network approaches have recently been applied [4, 47–49] for face recognition due to its simplicity, accuracy, generality, and efficiency in solving pattern recognition problems. The advantage of using neural networks for face recognition is the feasibility of a training system to capture the complicated class conditional density of face patterns [39, 50]. However, one drawback is that the network architecture has to be extensively tuned (number of layers, number of nodes, weight initialization, learning rates etc) to get exceptional performance. In principle, a neural network may be trained to recognize images directly. But for even a moderate size image, however the number of inputs will be very high. In such case, the network will be very complex and therefore will be difficult to train. To reduce the network complexity, neural network is often applied for pattern classification after feature extraction [13, 14, 51, 52]. It not only decreases the computational complexity of neural network but also increases the recognition rate.

The literature survey into the neural network structures, selected for recognition, broadly classifies them into two categories. In first structure, one single network is designed to classify all the classes; but in second structure, a single network is dedicated to recognize only one particular class, thus for each class there is a network. This book introduces a third category of structure for recognition using the higher-order neurons in complex domain where a single neuron is dedicated to recognize one particular class.

7.1.2.3 ACONN: All-Classes-in-One-Neural Network

An extensive survey into the litrature of the neural network tool for pattern classification point out the fact that the first architecture in a way as to ape the network arrangement in the classifier is one single neural network having as many outputs as the number of classes in the database. Thus, first structure viz. All-Classes-in-One-Neural Network (ACONN) is designed to classify all the classes (subjects) in a database through single huge network.

7.1.2.4 OCONN: One-Class-in-One-Neural Network

In the second architecture as a tool for pattern classification, an ensemble of neural networks is used, where each network is dedicated to recognize one particular class in the database. Thus, the second structure viz One-Class-in-One-Neural Network (OCONN), there are as many neural networks as the number of classes in the database. It has been experimentally verified in the literature [50, 52, 53] that the second classifier structure achieves better performance in comparison to the first structure. It leads to flexible structure of classifier, as a new network may easily be added when a new subject arrives in the database for recognition.

> Neural networks in a complex domain are becoming very attractive for solving many engineering problems. For a problem of same complexity, one needs a smaller network topology and lesser training time to yield better accuracy in comparison to equivalent neural networks in a real domain. Various experiments presented in this chapter on face recognition explore the computational power of the neurons in complex domain presented in Chap. 4.

7.1.2.5 OCON: One-Class-in-One-Neuron

The pattern recognizer (classifier) presented in this chapter is designed with an ensemble of higher-order neurons instead of ensemble of neural network. Therefore, each such neuron in complex domain is sufficiently powerful so that can take the responsibility for the recognition of individual subject of the database, hence lead a new kind of classifier, named as One-Class-in-One-Neuron (OCON). In this third architecture as a tool for pattern classification, each neuron is dedicated to recognize one particular class in the database. It lead to most compact but flexible structure of classifier where a new neuron may easily be added as a new subject arrives in the database for recognition.

So far, as implementation is concerned, a neuron in OCON or a network in OCONN is created with all the training samples of that class as positive example, termed as class-one; and the samples of other class as negative example, constitute the class-two. Therefore, recognition problem is a two class partitioning problem. The structure of each component (neuron or network) of classifier remains same for all classes and only weights wary. The weights of trained classifier are kept in a file and identification of input image is made on the basis of feature vectors and stored weights applied to each corresponding component one by one, for all the classes (subjects) in the test database.

7.2 Multivariate Statistical Techniques in Real and Complex Domain

A fundamental problem in digital image processing is to find a suitable representation for image, audio-video or other kind of data. Statistical methods like PCA-ICA find a set of basis images (Figs. 7.2 and 7.4) and represent images as a linear combination of these basis images. This reduces the amount of training image data prior to classification and provide reduction in computational cost while maintaining sufficient information for meaningful analysis. In most of the applications, an image, x, is available in the form of single or multiple views of 2D (p by q) intensity data (i.e., pixel values). Thus, inputs to the face recognition system are visuals only.

$$x = \{a_k : k \in S\}, \quad \text{where } S \text{ is a square lattice.} \tag{7.1}$$

Sometimes, it is more convenient to express an image matrix as a one dimensional vector of concatenated rows of pixels, then an image vector

$$x = \{a_1, a_2, ...a_N\}, \quad \text{where } N = p \times q \text{ is the total number of pixels.} \tag{7.2}$$

Such a high dimensional image space is usually inefficient and lacks discriminative power. Therefore, we need to transform 'x' into a feature vector or a new representation which greatly reduces the image feature dimensions, and yet maintains reasonable discriminative power by maximizing the spread of different faces within the image subspace. Let $X = \{x_1, x_2, ...x_M\}^T$ be the M by N matrix of image data, M is the number of images in the training set. Let vector $avg = \frac{1}{M}\sum_{k=1}^{M} x_k$ be the mean of training images. The goal in following feature extraction techniques is to find a useful representation of an image by minimizing the statistical dependence among the basis vectors.

7.2.1 Feature Extraction with RPCA

Principal component analysis in real domain (***RPCA***) is widely applied for dimensionality reduction and feature extraction by extracting the preferred number of principal components of multivariate data [21, 24, 26]. It captures as much variation as possible of training data set with fewer number of eigen vector subspace as possible. If image elements are considered to be random variables and images are seen as a sample of a stochastic process, the PCA basis vectors (eigenfaces) are defined as the orthonormal eigenvectors of the covariance matrix of image vectors. They describe a set of axis within the image space and most of the variances of images are along these axis. The eigenvalues associated with each eigen vector define the degree of spread (variance) of the image population in the direction of that eigenvector (along these axis). The eigenvectors associated with largest eigenvalue is the axis of maximum variance. The eigenvectors associated with second largest eigenvalue is the orthogonal axis, with second largest variance. Thus, the eigenvectors corresponding to higher eigenvalues carry significant information for representation, which best accounts for the distribution of images within the entire image space and so on. For most applications, the eigen vectors corresponding to very small eigenvalues are considered as noise and not taken into account during identification. All images in the training set are projected onto reduced subspace (set of eigenvectors corresponding to higher eigenvalues) to find a set of weights (new representation of images) that describes the contribution of each vector in the image space. By comparing the weights of the test image with the set of weights of the images in training set, the image in the test data can be identified.

Basic steps in ***RPCA*** algorithm for feature extraction can be summarized as follows:

1. Collect the images in data matrix X (M by N). Find the mean subtracted data matrix, $A = X - avg$.
2. Calculate the covariance matrix $C = A^T A$; where A^T denotes the transpose of A.
3. Find the eigenvectors (basis images) $A^T v_i$ (N by M) of C, such that $A^T A A^T v_i = e_i A^T v_i$.
4. In finding the eigenvectors of C, however for the moderate size of an image ($N = p \times q$), the dimension of C will be $pq \times pq$. Hence, calculations become quite unmanageable.
5. Turk [24] then circumnavigated the problem by proposing following scheme.
    - Consider the eigenvectors v_i of AA^T, such that

$$AA^T v_i = e_i v_i \tag{7.3}$$

    Each eigenvector v_i will be of size M and there will be such M eigenvectors. Thus, the calculations are greatly reduced from the order of number of pixels (N) in the image to the order of number of images (M) in the training set, $M \ll N$.
    - Multiply Eq. 7.3 both side by A^T, hence

$$A^T AA^T v_i = e_i A^T v_i \tag{7.4}$$

$$CA^T v_i = e_i A^T v_i \tag{7.5}$$

    The eigenvectors (basis images) of covariance matrix C will be $A^T v_i$ of size N by M. Thus eigenvectors of $A^T A$ (a N by N matrix) can be found from the eigenvectors of AA^T (a M by M matrix), which is easier to obtain, since generally $M \ll N$.

6. PCA has property of packing the energy into least number of principal components. The associated eigenvalues will be used to rank the eigenvectors according to their usefulness in characterizing the variation among the images. The eigenvectors (PCs) corresponding to the higher eigenvalues (i.e., subspace of basis images of size M′) carry significant information for representation.
7. Select the $M' < M$ eigenvectors (transformation matrix) from basis images corresponding to highest eigenvalues to extract facial features efficiently.
8. The new representation of an image is computed by projecting that image (input vector) onto subspace of basis images. After applying the projection, the input vector in a N dimensional space is reduced to a feature vector in a M′ dimensional subspace.

7.2.2 *Feature Extraction with* CPCA

The presented PCA in complex domain (***CPCA***) for feature extraction of image data is a generalization of PCA to complex variables. The complex ones are formed from the original data and their Hilbert transform. In Hilbert transformation, the amplitude of each spectral component is unchanged, but each component's phase is advanced by $\pi/2$ [12]. Complex principal components are determined from the complex cross-correlation matrix of image data matrix Z. Basic idea of ***C***PCA presented in [8, 42] has been used to factor the image matrix into a set of orthonormal basis vectors.

7.2.2.1 Algorithm

The well-known conditions of PCA or ***RPCA*** technique for extracting features has been extended to the complex domain, ie ***CPCA***. Basic steps in ***CPCA*** algorithm for feature extraction can be stated as follows:

1. Collect the images in data matrix X (M by N). Find the mean subtracted data matrix, $A = X - avg$.
2. Determine a complex image data matrix Z, using Hilbert transformation.
3. Compute the cross-correlation matrix $C = \mathrm{Z}^{\aleph}Z$, where $\mathrm{Z}^{\aleph}$ denotes the complex conjugate transposition of Z.
4. Find the eigenvectors of C. But for the moderate size of an image ($N = p \times q$), the dimension of C will be $pq \times pq$. Hence, calculations will be computationally extensive and intractable.
5. The problem can be circumnavigated by considering the eigenvectors v_i of $ZZ^{\aleph}$, such that $ZZ^{\aleph}v_i = e_i v_i$. Vector v_i is of size M and there are M eigenvectors. The calculations are greatly reduced from the order of number of pixels (N) in the image to the order of number of images (M) in the training set, $M \ll N$.

6. Eigenvectors (basis images) from the cross-correlation matrix C will be $Z^{\aleph}v_i$ (N by M), such that $Z^{\aleph}ZZ^{\aleph}v_i = e_i Z^{\aleph}v_i$.
7. Since, C is a hermitian matrix, its all eigenvalues will be real. In high dimensional image space, the energy mainly locates in the subspace constituted by first few eigenvectors. Thus, a significant compression can be achieved by letting those eigenvectors with large eigenvalues. Subspace capture as much variation as possible of training a set with less numbers ($M' < M$) of eigenfaces as possible.
8. Project each input vector (face image) onto basis vectors to find a set of M' coefficients that describe the contribution of each vector in the subspace.
9. Thus, input vector of size N is reduced to a new representation (feature vector) of size M', which be used for classification.

7.2.3 Independent Component Analysis

Independent component analysis (ICA) in real domain has emerged as a powerful solution to blind source separation and feature extraction problem. PCA aims at the decomposition of a linear mixture of independent source signals into uncorrelated components using only second-order statistics. Uncorrelatedness is a weaker form of independence. Basis images found by PCA separates only on pairwise relationships between pixels in the image database, however, higher-order relationships still appear in the joint distribution of basis images (PCA coefficients). In typical pattern classification problems, the significant information may be contained in the higher-order relationships among pixels. So, it will be desirable to use methods which are sensitive to these higher-order statistics to get better basis images. The assumption of Gaussian sources is implicit in PCA which makes it inadequate, because in real world, the data often does not follow a Gaussian distribution. ICA in real domain (ICA or ***R***ICA) is a method for transforming multidimensional random vectors into its components that are both statistically independent and non-Gaussian [35, 37, 54]. An important principle of ICA estimation is the maximization of the nongaussianity of linear combination of the observed mixture variables, which will yield the independent components. This transformation brings out the essential features of image data more visible or accessible. The ICA is the natural way to 'fine tune' the PCA. In researches it is observed that ICA defined subspace encode more information about image/data identity than PCA defined subspaces. The feature vectors obtained from ***R***ICA are as independent as possible and do not contain redundant data. Therefore, it is expected that these data of reduced dimension will be rich in features. ICA yields far better distinctiveness among classes of objects.

The feature extraction problem with one of the architectures of ICA or ***R***ICA [33, 38] can be stated as: Given a set of training images X (M by N), where images are random variables and pixels are number of observations. The image data consists of M variables (images) that have been observed together and the number of observations denotes the dimensionality of the image. ***R***ICA determines independent rather than uncorrelated image decomposition, thus provides a more powerful data representation than PCA. The idea is to find transformed images such that the resultant transformed vectors are statistically independent of each other. The IC representation of a face image is estimated on the basis of a set of spatially independent basis images obtained by ICA. This compressed representation of a face image is a vector of coefficients and may be used for linearly combining the independent basis images to generate the face image [38].

Start with assumption that the face images in X to be a linear mixture of statistically independent basis images S, combined by an unknown mixing matrix A, such that X = AS. ICA tries to find out the separating matrix W such that U = WX. ICA, as an unsupervised learning algorithm learns weight matrix W, which is used to estimate a set of independent basis images in the rows of U. W is roughly the inverse matrix of A. The ICA is carried out in a compressed and whitened principal components analysis space, where vectors associated small trailing eigenvalues are discarded. Therefore, before performing ICA, the problem of estimation of W can be simplified by two preprocessing steps:

- *Compression* Apply PCA on the image matrix X, to reduce the number of data to a tractable number, hence reduce the computational complexity. It also makes method convenient for calculating the representations of considered images. Preapplying PCA does not throw away the higher-order relationships, they still exist but not separated. It enhances ICA performance by discarding the eigenvectors corresponding to trailing eigenvalues, which tend to capture noise. Empirically it is seen that including the eigenvectors associated with small eigenvalues will lead to decreased ICA performance, as they amplify the effect of noise. Let matrix E^T (M′ by N) contains first M′ eigenvectors associated with higher eigenvalues of M face images in its row, then $U = WE^T$.
- *Whitening* Although PCA already removed the covariances in the data but variances were not equalized. Therefore, we need to retain the sphering step. Whitening process transforms the observed vector into a new vector, whose components are uncorrelated and variances are equalized. The whitening matrix is obtained as $W_w = 2 \times (COV(X))^{-(1/2)}$. The full transformation matrix will be the product of whiten (sphere) matrix and matrix learned by ICA. Instead of performing ICA on the original images, it should be carried out in a compressed and whitened space, where most of the representative information is preserved.

ICA observed wider interest and growing attention after 1990s when various approaches for ICA techniques were established. Among various prominent techniques for capturing independent components [33, 34, 37, 41, 54, 55], there are their own strength and weakness. One of the techniques for ICA estimation, inspired by information theory (infomax), is minimization of mutual information between random variables. This chapter formulates a complex version of ICA (i.e., *CICA*) for machine recognition using Bell and Sejnowski infomax method in real domain [38, 41]. Infomax performs source separation (to derive independent components) by minimizing the mutual information expressed as a function of higher-order cumulants. The algorithm for *CICA* has been derived from the principle of optimal information transfer through complex-valued neurons operated on nonanalytic but bounded activation function [56]. The main emphasis is to compare existing ICA or ***R*** ICA algorithm with ***C*** ICA for feature extraction in image database. Before summarizing infomax in real and complex domain it is necessary to briefly examine its derivation.

Mutual information is a natural measure of independence between random variables (sources). Infomax algorithm [35, 41] uses it as a criterion for finding the ICA transformation. All common ICA algorithms iteratively optimize a smooth function (Entropy), whose global optima occurs when the output vectors ($u \in U$) are independent. Finding independent signals by maximizing entropy is known as infomax [41].

Given a set of signal mixture X and a set of identical model cumulative distribution function (cdf) f of source signals (independent components). ICA works by adjusting the unmixing coefficient W, in order to maximize the joint entropy of the distribution of $Y = f(U = WX)$ or equivalently stating that probability density function (pdf) of Y is uniform. When W is optimal or Y has maximum entropy then extracted signals or basis images in U are independent.

7.2.4 Entropy and ICA Gradient Ascent

Entropy is defined as an average amount of certainty associated with a series of events. Less certainty corresponds to higher entropy hence uniform distribution of variables. For a set of signal mixture X consider the entropy of vector variable $Y = f(U)$ where $U = WX$ is the set of signals extracted by the unmixing matrix W. For a finite set of values sampled from a distribution with joint pdf $p_Y(Y)$ of Y, the entropy of Y can be defined as

$$H(Y) = -\int p_Y(Y) \ln p_Y(Y)\, dY \tag{7.6}$$

The joint entropy is essentially a measure of uniformity of a multivariate pdf $p_Y(Y)$ of Y. Let $S = (s_1, s_2, \ldots s_M)^T$ be matrix of unknown basis images (source signals). They have common model cdf f and model pdf p_s ($p_s = f'$). If joint entropy of $Y = f(U)$ is maximum then pdf p_u of each extracted signal/image in U will match the pdf p_s of source signal/images. If an unmixing matrix W exists which extracts signals/images in U from a set of image mixture X, then entropy of the Y is given as:

$$H(Y) = \left\langle \sum_{i=1}^{M} \ln p_s(U_i) \right\rangle + \ln|W| - \left\langle \sum_{i=1}^{M} \ln p_X(X_i) \right\rangle \tag{7.7}$$

where $\langle . \rangle$ stands for expected value. The last term in Eq. 7.7 is the entropy H(X) of the set of mixtures x. The unmixing matrix W that maximize H(Y) does so irrespective of the pdf $p_X(X)$, because $p_X(X)$ defines the entropy H(X) of the mixtures of x which can not be effected by W. Thus from general definition one may deduce that the last term (H(X)) in Eq. 7.7 can be ignored when seeking for an optimal W that maximizes equation Eq. 7.7, hence for optimization the entropy of the Y can be simplified as:

$$H(Y) = \left\langle \sum_{i=1}^{M} \ln\ p_s(U_i) \right\rangle + \ln|W| \tag{7.8}$$

By the definition, the pdf p_s of a variable is the derivative of that variable's cdf f, where f' is the pdf of every source signal. Thus,

$$p_s(U_i) = df(U_i)/dU_i$$

this derivative is denoted by $f'(U_i) = p_s(U_i)$; thus above equation can be rewritten as:

$$h(Y) = \left\langle \sum_{i=1}^{M} \ln\ f'(U_i) \right\rangle + \ln|W| \tag{7.9}$$

After obtaining the proper description of entropy in terms of the recovered signals and the unmixing matrix W, one need a method for finding that W which maximizes entropy of Y, and which therefore maximize the independence of U. The infomax algorithm for ICA given by Bell and Sejnowski (1995) states that "if the model pdf p_s matches the pdf p_u of the extracted signals then maximizing the joint entropy of Y also maximize the amount of mutual information between X and Y".

The estimation of optimal W can be achieved by performing gradient ascent on the entropy of the output Y with respect to weight matrix W. The gradient ascent can effectively be performed on h(Y) by iteratively adjusting W as in neural network (in order to maximize the function h(Y)). One can start by finding the partial derivative of h(Y) with respect to one scalar element W_{ij} of W. The weight W_{ij} determines the proportion[1] of the jth mixture in the ith extracted signal U_i. Given that $U = WX$ and that every source signal has the same pdf f', the partial derivative of h(Y) with respect to the ijth element in W is

$$\frac{\partial h(Y)}{\partial W_{ij}} = \left\langle \sum_{i=1}^{M} \frac{\partial \ln f'(U_i)}{\partial W_{ij}} \right\rangle + \frac{\partial \ln|W|}{\partial W_{ij}} \tag{7.10}$$

The interested readers may easily evaluate each of the two derivatives on the right hand side of Eq. 7.10 in turn. Further simplification and rearrangements of terms yields:

$$\frac{\partial h(Y)}{\partial W_{ij}} = \left[W^T\right]_{ij}^{-1} + \left\langle \sum_{i=1}^{M} \frac{f''(U_i)}{f'(U_i)} x_j \right\rangle \tag{7.11}$$

[1] Precisely talking, W_{ij} ascertains the proportion only if the weights that contribute to U_i sum to unity, this is of no significance for our objectives.

$\left[W^T\right]_{ij}^{-1}$ is the *ij*th element of the inverse of the transposed unmixing matrix W. The gradient (∇) of h (i.e., ∇h) is the matrix of derivatives in which the *ij*th element is $\partial h/\partial W_{ij}$. If we consider all the elements of W then

$$\frac{\partial h(Y)}{\partial W} = \left[W^T\right]^{-1} + \left\langle \frac{f''(U)}{f'(U)} X^T \right\rangle \tag{7.12}$$

Given the finite samples of M observed mixture values in X^T and a putative unmixing matrix W, the expectation can be estimated as the mean

$$\nabla h = \left[W^T\right]^{-1} + \frac{1}{M} \sum_{i=1}^{M} \frac{f''(U_i)}{f'(U_i)} X^T \tag{7.13}$$

Therefore, in order to maximize the entropy of $Y = f(U)$, the rule for updating W according to gradient ascent on joint entropy ($\Delta W \propto \partial\, h(Y)/\partial W$) comes out in its most general form as follows:

$$\Delta W = \eta \left(\left[W^T\right]^{-1} + \frac{1}{M} \sum_{i=1}^{M} \frac{f''(U_i)}{f'(U_i)} X^T \right) \tag{7.14}$$

where η is the learning rate. One can easily drive the expression for ∇h for a specific cdf of the source signals:

- A commonly used cdf to extract source signals is the logistic function. If logistic nonlinear function (Fig. 2.1) is used then rule for updating W could be obtained using Eq. 7.14 by replacing:

$$\frac{f''(U_i)}{f'(U_i)} = 1 - 2Y \tag{7.15}$$

- Another important model cdf for extracting the super-gaussian (high kurtosis) source signals is the hyperbolic tangent function. Given the cdf $f(U) = \tanh(U)$ then the gradient ascent learning rule for updating W could be obtained using Eq. 7.14 by replacing:

$$\frac{f''(U_i)}{f'(U_i)} = -2\tanh(U) \tag{7.16}$$

The Infomax Algorithm evaluates the quality of any putative unmixing matrix W using Eq. 7.14 through given set of observed mixtures X and corresponding set of extracted signals U. Thus, one can deduce that, for the optimal unmixing matrix, the signals $Y = f(U)$ have maximum entropy and therefore independent. If f is chosen as the model cdf of source signals then maximization of the entropy of neural network output is equivalent to minimization of mutual information between the individual outputs in U [40, 41]. A single layer neural network set up, which implement $Y = f(WX)$, will be used for weight updation according to gradient ascent on joint entropy ($\Delta W \propto \partial h(Y)/\partial W$). The nonlinearity f in neuron should roughly approximate the cdf of source distribution. Therefore, presented infomax algorithm in real and complex domain proceed by maximizing the entropy of output of a single-layered neural network.

7.2.5 Feature Extraction with RICA

Let X be the matrix of observed variables and U is its transformation given by some matrix of coefficients W, $U = WX$. The ICA may be stated as finding the matrix W so that the random variables $u_i \in U$ (rows of U) are as independent as possible (finding statistically independent basis images). The goal of presented ICA algorithm is to maximize the mutual information between the environment X and the output of the neural network Y. This is achieved through gradient ascent on the entropy of the output with respect to the weight matrix W. Basic steps in ***RICA*** algorithm derived from the principle of optimal information transfer in neurons with sigmoidal transfer functions, for feature extraction, can be summarized as follows:

1. Collect the images in data matrix X (M by N) so that images are in rows and pixels are in column.
2. Apply ***RPCA***. The PCA basis vectors in E^T are analyzed for independent components, where E (N by M′) be the matrix of M′ eigenvectors of M images.
3. Apply whitening (sphering) of data. Final transformation matrix will be the product of whitening matrix and optimal unmixing matrix.
4. Sources are modeled as real random vectors. Take sigmoidal transfer function f as joint cdf of source signals, in view of optimal information transfer in real-valued neural network.
5. Derive a contrast function (joint entropy 'h') in view of real-valued neural network. Perform maximization of entropy $h(Y)$ by neural network using gradient ascent. If X is the input vector to ANN then $f\,(WX)$ is the output.
6. Find the optimal matrix W such that: MAX $[h\,\{f\,(WX)\}]$, this can be done as:
   - Define a surface $h\,\{f\,(U)\}$
   - Find the gradient ∇h with respect to W and ascent it, $\Delta W \propto \nabla h$, then

$$\Delta W = \eta\left(\left[W^T\right]^{-1} + \frac{1}{M}\sum_{i=1}^{M}\frac{f''(u_i)}{f'(u_i)}X^T\right) \quad u_i \in U. \tag{7.17}$$

   Where, $f_C''(u)/f_C'(u) = (1 - 2\,f(u))$.
   - When h is maximum, W is W_{OPT}. done!

   The ICA algorithm learns weight matrix W, which is used to estimate a set of independent basis images in the rows of U. Projecting the eigenvectors onto learned weight vectors produces the independent basis images.

7. Maximizing joint entropy of output Y gives independent basis images in U. It also minimize the mutual information between the individual outputs (basis images). Thus, ICA algorithm produces transformation matrix $W_t = W_{OPT} \times W_w$ such that $U = W_t\, E^T$. That will let us know how much the extracted signals in U are close to being independent.
8. Let R be the M by M′ matrix of PC representation of the images in X, R = XE, also approximation of X = RE^T. Assumption that W is invertible, we get $E^T = W_t^{-1} U$. Hence $X = RW_t^{-1} U$.
9. Each row of matrix $B = RW_t^{-1}$ contains the coefficients for the linear combination of statistically independent basis images in U; $X = BU$. This X comprises the images in its rows, X is the reconstruction of the original data. Thus, statistically independent feature vectors (IC representation) of images have been obtained.

7.2.6 Feature Extraction with **CICA**

Independent component analysis in complex domain ($CICA$) has been used for source separation of complex-valued data such as fMRI [10], EEG [9] and communication data, yet this concept is not well developed hence demanding more applications. The most important application of $CICA$, in machine recognition is still untouched, therefore it is worthwhile to develop the feature extraction algorithm with basic concepts of ICA. This chapter is devoted to build $CICA$ algorithm for image processing and vision applications. It is observed in Chap. 4 that ANN in a complex domain gives a far better performance in the real-valued problems. It will be fruitful to investigate the principle of optimal information transfer through complex-valued neurons, incorporating nonanalytic activation function, for feature extraction from image database. A bounded but nonanalytic activation function f_C given in Eq. (3.3) has performed well with neurons in complex domain [56]. The motivation of using this function in CICA algorithm is that it can approximate roughly well the joint cdf ($F(u_{\Re}, u_{\Im})$) of source distribution [57, 58]. The apparent problem in this complex function comes from the fact that it is real-valued and therefore is not complex differentiable unless it is a constant. The differentiation of f_C can be conveniently done [59, 60] without separating real and imaginary parts with following complex differential (partial) operator:

$$f_C'(z) = \frac{\partial f_C(z)}{\partial z} = \frac{1}{2}\left(\frac{\partial f_C(z)}{\partial z_{\Re}} - j\,\frac{\partial f_C(z)}{\partial z_{\Im}}\right) \tag{7.18}$$

CICA algorithm is derived by maximizing the entropy of the output from a single layered complex-valued neural network. The update equations in unsupervised learning involve first- and second-order derivatives of the nonlinearity. The complex-valued sigmoid function is flexible enough to obey the joint cdf. The Infomax algorithm [9, 43, 58, 61] in complex domain setup is modified for feature extraction in this section using split nonlinearity. We perform maximization of entropy surface by complex gradient ascent which result in faster convergence. This is achieved through the complex extension of the natural gradient. In ***C***ICA-based feature extraction, we assume that the image data X is a linear mixture of M statistically independent complex-valued sources S, then Y = f_C (U = W X), where U, W, X $\in C$. The basic steps in ***C***ICA algorithm for feature extraction in image database can be summarized as follows:

1. Collect the images in data matrix X (M by N) so that images are in rows and pixels are in column.
2. Apply ***R***PCA. The PCA basis vectors in E^T are analyzed for independent components, where E (N by M′) be the matrix of M′ eigenvectors.
3. Apply whitening (sphering) of data. Final transformation matrix will be the product of whitening matrix and optimal unmixing matrix.
4. Sources are modeled as complex random vectors. Take sigmoidal complex function f_C, defined in Eq. (3.3), as joint cdf of source signals.
5. Derive a contrast function h from CVNN point of view. Perform maximization of joint entropy $h(Y)$. This can be achieved using extension of the natural gradient in a complex domain.
6. Perform complex infomax on PCA basis X. Find an optimal matrix W such that: MAX $[h\,\{f_C\,(WX)\}]$, this can be done as:

   - Define a surface $h\,\{f_C\,(U)\}$
   - Find the gradient ∇h with respect to W and ascent it, $\Delta W \propto \nabla h$, then

$$\Delta W = \eta\left(\left[W^T\right]^{-1} + \frac{1}{M}\sum_{i=1}^{M}\frac{f_C''(u)}{f_C'(u)}X^T\right), \quad u \in U. \tag{7.19}$$

$$\text{where } \frac{f_C''(u)}{f_C'(u)} = \frac{f'(\Re(u))(1-2\,f(\Re(u))) - jf'(\Im(u))(1-2f(\Im(u)))}{2(f'(\Re(u)) + f'(\Im(u)))}. \tag{7.20}$$

   - When function h is maximum or magnitude of gradient of function h converges toward zero, W is W_{OPT}. done! At a maximum in h the gradient magnitude should be zero.

7. Maximizing joint entropy of outputs also minimizes the mutual information between the individual outputs, ie basis images in U. That will let us know how much the extracted signals in U are close to being independent. Thus, ***C***ICA algorithm produces transformation matrix $W_t = W_{OPT} \times W_w$, such that $U = W_t E^T$.
8. Let R = XE be the PC representation of images in X, also approximation of X = RE^T. Assumption that W_t is invertible, we get $E^T = W_t^{-1}U$. Hence $X = RW_t^{-1}U$. The estimation of IC representation of images is therefore based on the independent basis images in U.

9. Each row of matrix $B = RW_t^{-1}$ is IC representation of a image. Thus, statistically independent feature vectors of images have been obtained in B. ***R***PCA as preprocessor also ensures that ***C***ICA algorithm does not magnify the effects of noise.

7.3 Human Recognition Systems

The machine recognition is both important and challenging technique. This chapter is aimed at developing an intelligent machine recognition system that is noise invariant and can work well in occluded and blurred environment. This also aims at reducing the computational cost and provides a faster recognition system with complex-valued neurons presented in Chap. 4. This chapter explores the abilities of real and complex ICA for feature extraction and compare them with real and complex PCA. In order to verify the utility of neurons-based classifiers in complex domain, the empirical studies are on two standard face data sets are presented. This chapter demonstrate the comprehensive performance of various neuron models along with feature extraction techniques in terms of training epochs for learning, number of learning parameters (weights) for storage and accuracy of recognition system in varying conditions.

Recent developments in machine intelligence and call for better security applications have brought biometrics into focus. It is known that signature, handwriting, voice, and fingerprint have a long history. More recently retinal scan, iris scan, periocular, occulomotion and facial information are considered for biometrics. When deploying a biometrics based system, we consider its accuracy, cost, ease of use, whether it allows integration with other systems as well as the ethical consequences of its use. Biometrics can be defined as the automated use of physiological and behavioral characteristics for verifying or recognizing the identity of living person. The example involved in the measurement of biometric features are classified as per their characteristics:

1. Physiological features: face, fingerprints, palm print, DNA, iris, periocular, hand geometry etc.
2. Behavioral features: signature and typing rhythms, voice, gait, etc.

Physiological features are more stable than behavioral. The reason is that physiological features are often nonalterable except in the case of severe injury, while behavioral features may fluctuate due to stress, fatigue, or illness. The interest and research activities in security systems have been increased over the past few years. This growth is largely driven by a growing demand for identification and access control at very important places. For the purpose of better security applications, the variety of identification attributes can be classified into three broad categories:

- Lowest level security: something we have, such as photo ID.
- Middle level security: something we know, such as password or PIN.
- Highest level security: something we do or something we are, which comprises physiological and behavioral biometrics.

Face recognition is one of the few biometric methods that possesses the merits of both high accuracy and low intrusiveness. It is also one of the most acceptable biometrics because a human face is always bare and often used for its visual interactions. It cannot be forgotten or mislaid like a password and it has the potential to characterize a person without any document for identification. The success of intelligent face recognition system mainly depends on the development of a computational model for facial feature representation and classifiers for identification of an unknown person.

7.4 Recognition with Higher-Order Neurons in Complex Domain

Classification needs to be performed on mathematical models (extracted features) given by aforementioned statistical techniques. A neural network is an artificial intelligence technique, which has many advantages for nonlinear classification because of fast learning, better generalization, efficiency and robustness toward noise and natural environment [50, 51]. In case of learned networks, the weights can be easily communicated to humans for generalization than learned rules (in traditional AI techniques). The efficient learning ability and incredible generalization of single RSP and RSS neurons in complex domain have motivated to use them in designing the OCON-based machine recognition system. In this chapter, an ensemble of complex-valued neurons have been used to develop a OCON classifier.

The training of ANN classifiers involves the estimation of learning parameters (weights) only, which are stored for future testing. Hence, it is most desirable to search a structure for classifier which require minimum weights and yield best accuracy. The beauty of proposed classifier structure for image classification system is that it uses an ensemble of proposed single neurons in complex domain instead of ensemble of multilayer network. Only single RSS or RSP neuron has given solution to many benchmark problems in previous chapters. Similar computational power is also observed in image classification, where every single neuron is dedicated to recognize images of its 'own class' assigned to it. The output of the ensemble is forming an aggregate output of the classifier.

The number of neurons in ensemble is set to the number of image classes (subjects). Each neuron is trained to give output '$1 + j$' for its own class and 0 for other class. The number of nodes in input layer is equal to the number of elements in the feature vector. The data values in the feature vectors are normalized to lie with in the first quadrant of a unit circle. An input face image presented to a neuron associated to its own class is considered as positive example, while images of other classes to this neuron is considered as negative example. It yields better confidence in classifier decision. The classifiers may be trained with error back-propagation learning algorithm. For testing the identity claim of an image, a M' dimensional feature vector is extracted from the image and this vector is given to every neuron or network of classifier. The discriminant function applied in the proposed system calculates the square deviation from actual output to the ideal output. An unknown feature vector is classified as belonging to a class, when deviation is less than a threshold. In general, the threshold can be determined from the experimental studies.

7.5 Performance Evaluation with Different Benchmark Database

In order to demonstrate the effectiveness of presented feature extractors and classifiers, two sets of experiments are demonstrated in this section for relative performance evaluation. In these experiments, the general performance of classifiers based on different neuron architectures are presented. The effect of using different statistical feature extraction techniques in real and complex domain is also demonstrated with different classifiers. In all experiments, four images of each subject are taken for training and remaining images are considered for testing. The performance analysis results presented in this chapter are average outcome of five training processes with different weight seedings.

In order to evaluate the performance of different techniques, three standard biometric measures are used. Recognition error rate (in other way percentage accuracy) is the number of misclassification divided by total number of test images. False acceptance rate (FAR) is the number of unauthorized persons considered as authorized divided by total number of unauthorized attempts. False rejection rate (FRR) is the number of authorized persons considered as unauthorized divided by total number of authorized attempts. FAR and FRR are inversely proportional measurements, therefore variable threshold setting are provided for users to keep balance. It is generally desirable that recognition system must reject unauthorized attempts as much as possible and perform well for authorized attempts. Thus, system must be much stricter to unauthorized persons and slightly stricter for authorized persons. In all experiments, our selection of threshold is inclined in a direction where we can obtain significantly lower FAR and slightly higher FRR.

7.5.1 Performance in ORL Face Database

The first database considered for evaluating performance of proposed technique is AT & T laboratories face database (formerly ORL) of Cambridge University [62]. It contains facial images of 40 different subjects (persons) and 10 images per subjects. None of the ten samples of each subjects is identical to each other. The images are taken against a dark homogeneous background. There are variations in facial expressions (open or closed eyes, smiling or frowning face), facial details (with or without glasses, hairs), scale (up to 10 %), orientation (upto 20°). Each image was digitized and presented by 92×112 pixel array whose gray level ranged between

0 and 255. As an example, few sample images are shown in Fig. 7.1. A total of 160 images (four of each subject) have been used for training and rest 240 for testing. Feature vector for each image has been created based on 48 dominant principal components, which are used into classification or recognition. The squared error or absolute error from each output is used to decide the identity claim in a subject. The claim is accepted if error is less than threshold.

Eigenvectors with higher eigenvalues provide more information on the face variation than those with smaller eigenvalues. The eigenvectors (eigenfaces) corresponding to largest eigenvalues, derived from the ORL face database, are shown in Fig. 7.2. Similarly, the eigenvectors corresponding to the smallest eigenvalues are shown in Fig. 7.3. The eigenvectors are ordered according to decreasing eigenvalues in each figure. Each individual face in the face set can then be approximated by a linear combination of the eigenvectors associated with largest eigenvalues. Similarly, independent components obtained by ICA algorithm are used as feature vectors for classification. ICA separates the high-order moments of the input in addition to the second-order moments utilized in PCA [31, 38]. The basis vectors obtained by Infomax algorithm for ICA representation are shown in Fig. 7.4.

Eigenvalues in ***RPCA*** and ***CPCA*** can be used to plot a graph (Fig. 7.5) of variance captured by each principal component. It is essentially the cumulative distribution for M' components and defined in Eq. (7.21). This graph allows us to select the dominant eigenvectors for feature extraction. It is found that 48 eigenvectors or 30 % of maximum possible number of eigenvalues (training set has 160 images) of *C*PCA are enough to account for more than 93 % of the variations among the training set. While, approximately 85 % of total variance is retained in same number

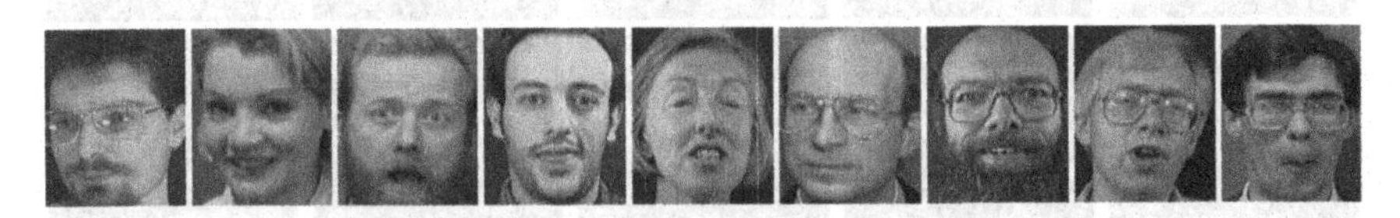

Fig. 7.1 Some example images from the ORL face database

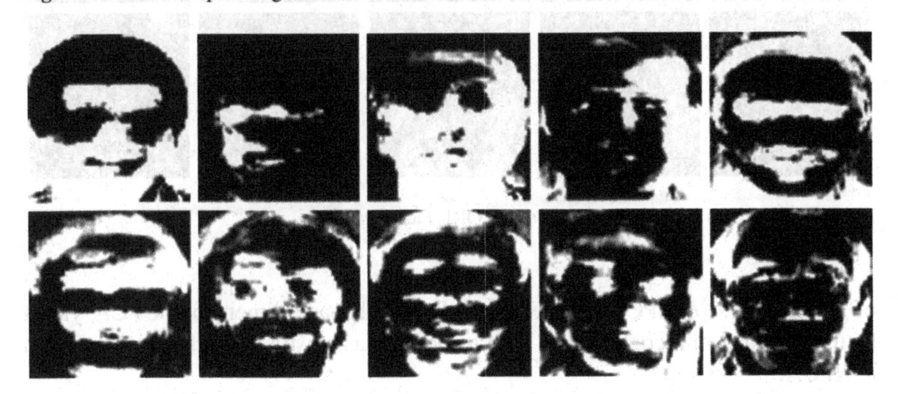

Fig. 7.2 Eigenvectors corresponding to the largest eigenvalues, derived from the ORL face database

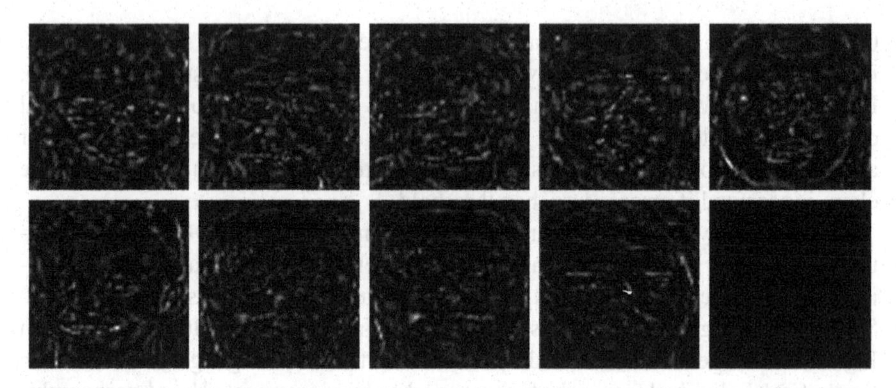

Fig. 7.3 Eigenvectors corresponding to the smallest eigenvalues, derived from the ORL face database

Fig. 7.4 ICA basis vectors derived from the ORL face database

of eigenvectors of ***R***PCA. Therefore with 48 subspace dimensions ***C***PCA yields better accuracy in recognition as well as better class distinctiveness in comparison to ***R***PCA.

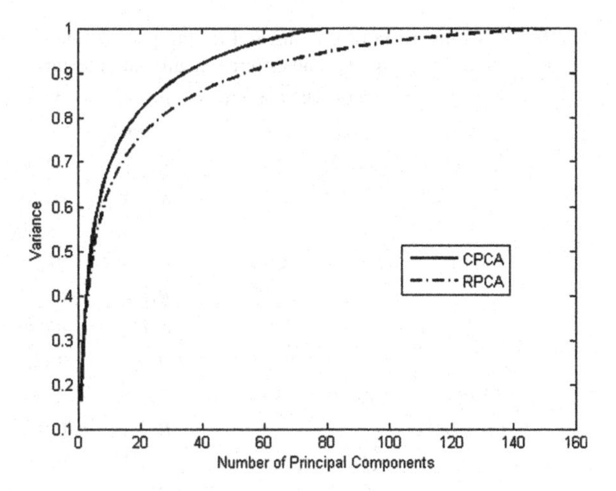

Fig. 7.5 Graph of variance captured by each principal component in the ORL face database

$$\frac{\sum_{i=1}^{M'} e_i}{\sum_{M}^{i=1} e_i} \quad \text{where} \quad M' < M \ll N. \tag{7.21}$$

Performance comparison of PCA- and ICA-based feature extraction techniques in real and complex domain along with different neurons-based classifier is presented in Table 7.1. Classifier with complex-valued neurons are fairly better than with real-valued neurons. Results clearly demonstrate that ***C***RSP and ***C***RSS neuron-based classifier yield best accuracy with smallest topology. They are far efficient in comparison with any other neurons. The results demonstrating the recognition performance in terms of different parameters such as training epochs, number of learning weights, FRR, FAR, recognition rate are presented in Table 7.1. The performance of different ANN architectures are broken down according to the feature extraction methods. The striking feature of Table 7.1 is that classifier-based on RSP and RSS neurons in complex domain requires least training epochs and learning parameters with highest recognition rate. The ***C***ICA-based feature extraction always provides the best results. Though, difference among different feature extraction methods in the context of accuracy is not large, but the significant thing is the class distinctiveness among different subjects. The class distinctiveness is much better in ***C***ICA-based system in comparison with any other method. Thus, in case of ***C***ICA, the deviation of square error from threshold is quiet large for class and nonclass faces, which makes it in turn robust to perform more effectively in real life situations where variety of noises and occlusion may be present in each image. From extensive simulations on ORL face database, it is imperative to make following inferences:

- Class distinctiveness is very poor in case of ***R***PCA and ***C***PCA.
- Class distinctiveness with ***R***ICA is slightly better than PCA but it is best with ***C***ICA.
- ***C***ICA-based system is much stricter to unauthorized person.
- ***C***RSP neuron-based classifier is outperforming over all other classifiers.

Table 7.1 Comparison of training and testing performance for ORL face dataset with different feature extraction techniques and different neuron architectures

S. No.	Neuron type type	Network	Parameters	Feature extraction	Average epochs	FRR	FAR	Recognition rate (%)
1	***R*MLP**	48-8-1	40 × 401	***RPCA***	44,000	0.092	0.032	96.6
		48-7-1	40 × 351	***RICA***	6,000	0.067	0.014	98.25
2	***C*MLP**	48-3-1	40 × 151	***RPCA***	28,000	0.083	0.020	97.8
				CPCA	28,000	0.092	0.018	98.0
				RICA	6,000	0.087	0.016	98.2
				CICA	6,000	0.067	0.0095	98.9
3	***C*RPN** ($d = 0.9$)	48-2-1	40 × 101	***RPCA***	28,000	0.080	0.018	98.0
				CPCA	28,000	0.080	0.016	98.2
				RICA	6,000	0.075	0.018	98.0
				CICA	6,000	0.070	0.011	98.75
4	***C*RSS**	48-1	40 × 99	***RPCA***	6,000	0.092	0.020	97.7
				CPCA	6,000	0.080	0.014	98.3
				RICA	4,000	0.083	0.017	98.1
				CICA	4,000	0.075	0.009	99.0
5	***C*RSP**	48-1	40 × 99	***RPCA***	6,000	0.086	0.020	97.85
				CPCA	6,000	0.075	0.009	98.8
				RICA	4,000	0.083	0.013	98.4
				CICA	4,000	0.070	0.006	99.25

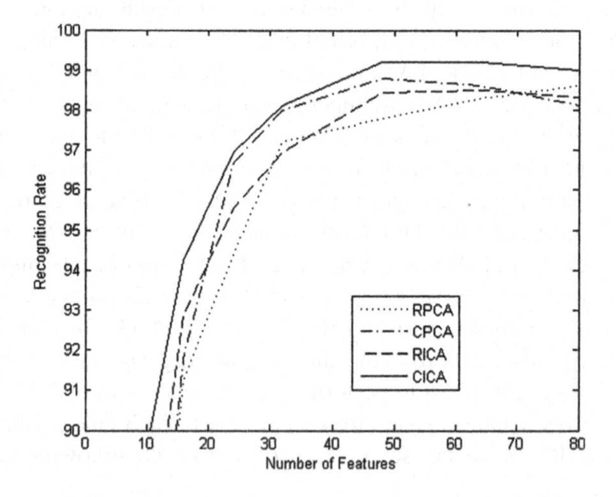

Fig. 7.6 Recognition rate versus subspace dimension for different feature extraction techniques in ORL face database

In order to assess the sensitivity of different feature extraction methods on the number of subspace dimension variations, the comparative performance is shown in Fig. 7.6. The results of experiments are typical in part, because the recognition rate does not increases monotonically with the number of subspace dimensions.

Figure 7.6 shows that the feature extraction criterion performs reasonably good, when 48 features of subspace are chosen. The increase in subspace dimension does not improve the overall performance. Though recognition rate in ***R***PCA increases slowly beyond 48 features, but still it is less than ***C***ICA.

7.5.2 Performance in Indian Face Database

This experiment considers 500 color images corresponding to 50 subjects of Indian face database [63]. Each image is of size 640×480 pixels. All images have a bright homogeneous background and different poses and emotions. For each individual, following poses have been included for the face: looking front, looking left, looking right, looking up, looking up toward left, looking up toward right, and looking down. In addition to the variation in pose, images with four emotions such as neutral, smile, laughter, and sad/disgust are also included for every individual. For example, few sample images are shown in Fig. 7.7. The number of images considered for training and testing is 200 and 300 respectively, comprising of all subjects.

Eigenvectors with higher eigenvalues provide more information on the face variation than those with smaller eigenvalues. The eigenvectors (eigenfaces) corresponding to largest eigenvalues, computed from the Indian face database are shown in Fig. 7.8. The eigenvectors are ordered according to decreasing eigenvalues. Each individual face in the face set can then be approximated by a linear combination of the eigenvectors associated with largest eigenvalues. Based on the argument that for tasks such as face recognition much of the important information is contained in high-order statistics; it has been proposed [31, 38] to use ICA to extract features for face recognition. The basis vectors (basis images) obtained by the Infomax algorithm for ICA representation are shown in Fig. 7.9. We obviously want to capture as much variations as possible of the training set with as fewer number of subspace dimensions. The graph in Fig. 7.10 allows us to see more clearly how much variation is captured by number of eigenvectors. This information prompted us to select 60 eigenvectors from the set of training images of 50 subjects for further feature extraction. They correspond for keeping more than 91 % of total variance in the eigenvalues of this dataset.

The test results of Indian face database with different feature extraction techniques and different neural classifiers are presented in Table 7.2 in terms of different measures. It again reveals the superiority of ***C***RSP-based classifier. The feature vectors

Fig. 7.7 Some example images from the Indian face database

Fig. 7.8 Eigenfaces computed from Indian face database

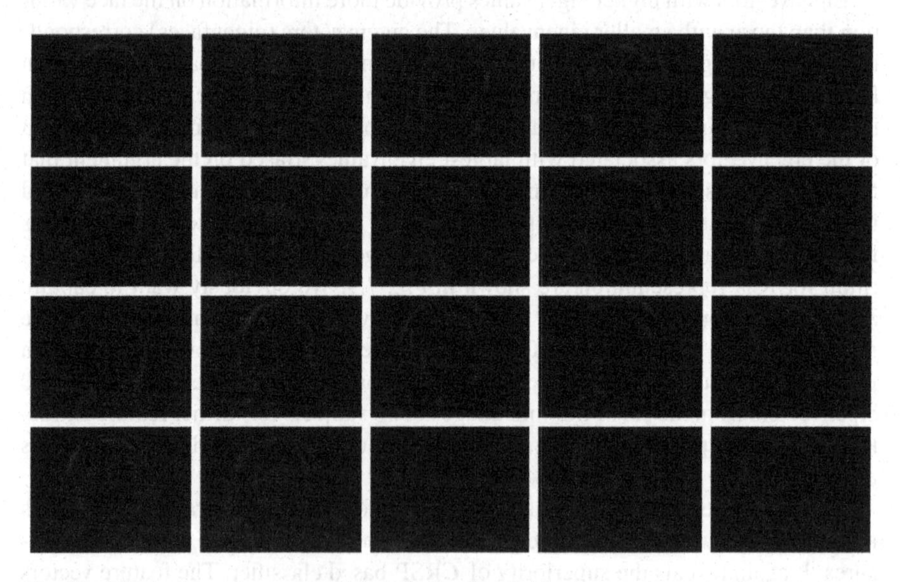

Fig. 7.9 ICA basis images computed from Indian face database

yielded by ICA-based methods provide faster learning in all neural classifiers. It is worthwhile to mention that this face dataset has much variations in poses, which decrease its recognition accuracy. But, performance of *CICA* in feature extraction is significant in terms of precision in the classification and accuracy. With *CICA*, classification error is quiet low for class faces while it is quiet high for nonclass faces.

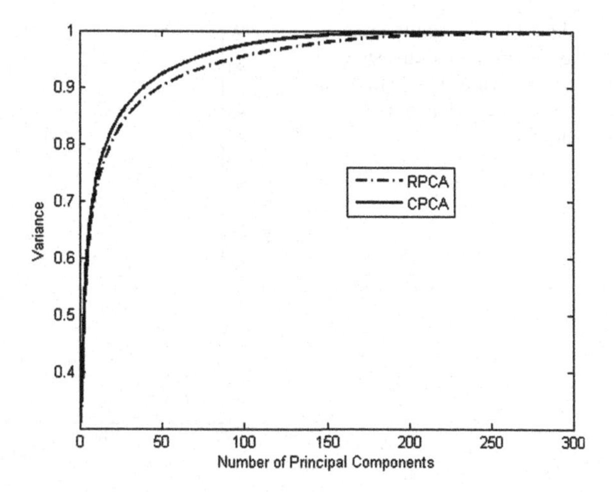

Fig. 7.10 Graph of variance captured by each principal component of Indian face database

Table 7.2 Comparison of training and testing performance for Indian face dataset with different feature extraction techniques and different neuron architectures

S.no.	Neuron type	Network	Parameters	Feature extraction	Average epochs	FRR	FAR	Recognition rate (%)
1	***R*MLP**	60-8-1	50 × 497	***RPCA***	45,000	0.19	0.055	93.8
				RICA	25,000	0.17	0.045	95.3
2	***C*MLP**	60-3-1	50 × 187	***RPCA***	30,000	0.20	0.044	95.0
				CPCA	30,000	0.21	0.046	94.8
				RICA	10,000	0.17	0.040	95.5
				CICA	10,000	0.17	0.035	96.6
3	***C*RPN** ($d = 0.94$)	60-2-1	50 × 125	***RPCA***	30,000	0.19	0.047	94.8
				CPCA	30,000	0.20	0.047	94.8
				RICA	10,000	0.19	0.045	95.2
				CICA	10,000	0.18	0.037	96.1
4	***C*RSS**	60-1	50 × 123	***RPCA***	20,000	0.17	0.042	95.3
				CPCA	20,000	0.18	0.046	95.0
				RICA	10,000	0.20	0.041	95.5
				CICA	10,000	0.16	0.029	96.8
5	***C*RSP**	60-1	50 × 123	***RPCA***	20,000	0.15	0.041	95.4
				CPCA	20,000	0.16	0.043	95.2
				RICA	10,000	0.20	0.037	95.9
				CICA	10,000	0.17	0.025	97.2

From extensive simulations on this face dataset, it is imperative to make following inferences:

- Class distinctiveness is very poor in case of ***RPCA*** and ***CPCA***.

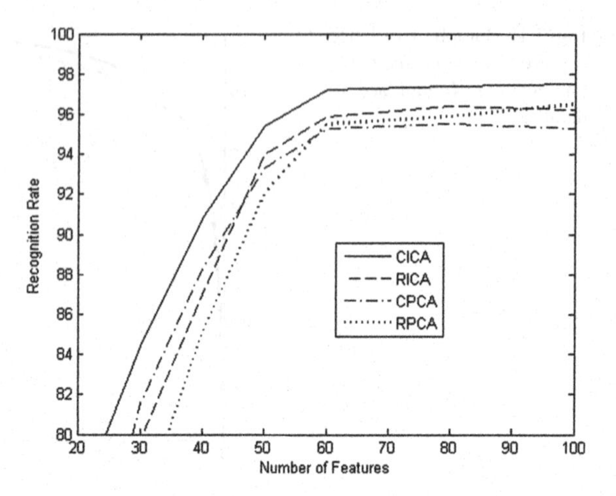

Fig. 7.11 Recognition rate versus subspace dimension for different feature extraction techniques in Indian face database

- Class distinctiveness with ***RICA*** is slightly better than PCA but it is best with *CICA*.
- The recognition system developed with ***C***ICA and ***C***RSP neuron is outperforming in all respect over others.

This makes ***CICA*** to perform very well in noisy and blurred images, which are generally captured in real environment applications. The comparative assessment of different feature extraction methods as a function of the dimension of compressed subspace is presented in Fig. 7.11. It is obtained from best performing ***C***RSP classifier. ***CICA*** outperforms to all other methods, it is statistically more significant when we consider a lesser number of subspace dimensions. The relative ordering of the subspace projection techniques depends on the number of subspace dimensions. The performance of ***CPCA*** is better in lower dimensions while ***RPCA*** is better in higher dimensions. It can be seen that the recognition rate with ***CICA*** is always higher than any other technique.

7.6 Robustness Toward Partial Occlusion, Blurring, and Noises

In order to examine the performance of different techniques in varying environmental conditions, the occlusion has been simulated at random locations in few images, as in Figs. 7.12 and 7.13. Some of the noises and blurring has also been introduced (electronically modified) [14, 34] in other images of face data base for experiments, as in Fig. 7.14. In this experiment, we consider the best performing classifier of ***C***RSP neurons. Both *PCA*-based face recognizers have fairly recognized the occluded faces shown in Fig. 7.12 where the degree of occlusion is quiet less. On increasing the degree of occlusion (Fig. 7.13), it is observed that the performance of both ***RPCA***

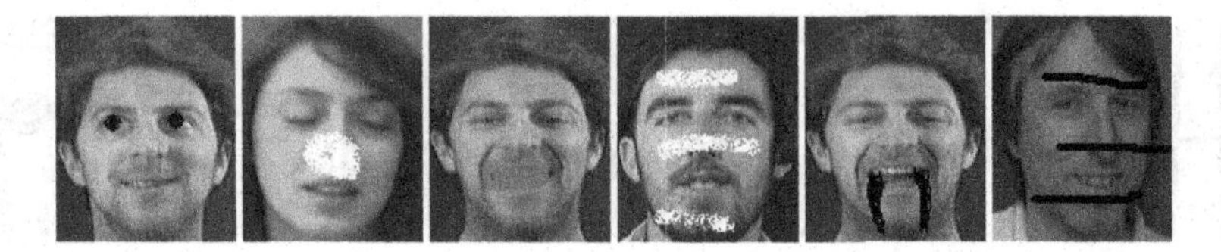

Fig. 7.12 Occluded images, recognized by both *PCA*-based recognition system as well by both *ICA*-based recognition system

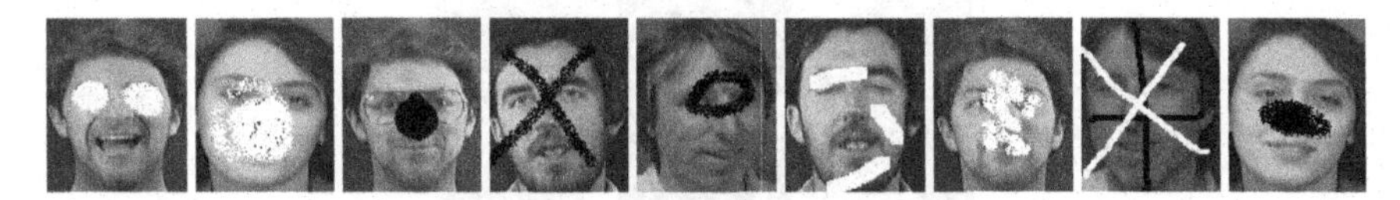

Fig. 7.13 Occluded images, not recognized by any *PCA*-based recognition system but correctly identified by both *ICA*-based recognition system

Fig. 7.14 Blurred images, not recognized with any *PCA*-based recognizer but correctly identified by both *ICA*-based recognition system

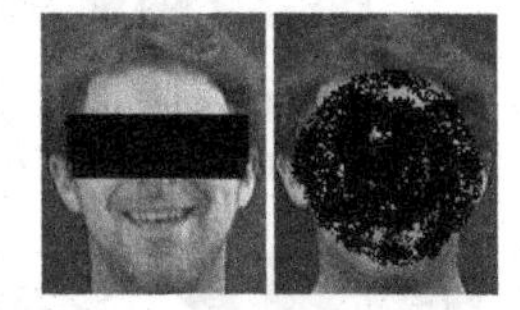

Fig. 7.15 Occluded images which are not recognized with any feature extraction technique

and ***CPCA*** is very poor. Set of faces in Fig. 7.13 were not recognized by any PCA based methods. Similarly, Fig. 7.14 presents a set of blurred images which were also not correctly identified by any PCA-based methods. The reason is poor class distinctiveness offered by both PCA-based feature extraction techniques. But, the set of images in Figs. 7.13 and 7.14, having comparatively more distortions, can be correctly classified by ***RICA*** and ***CICA***-based methods while any PCA-based method failed to do so.

Therefore, it is apparent that the recognition with different feature extraction techniques depends on the degree of occlusion and blurring introduced [34] in electronically modified images. Figures 7.15 and 7.17 present faces with very high degree of occlusion and blurring, respectively. The experiments found that these images can not be recognized by any technique. In another set of images, comparatively less degree of occlusion and blurring, as shown in Figs. 7.16 and 7.18, but comparatively more degree of occlusion and blurring than Figs. 7.13 and 7.14, respectively; it was

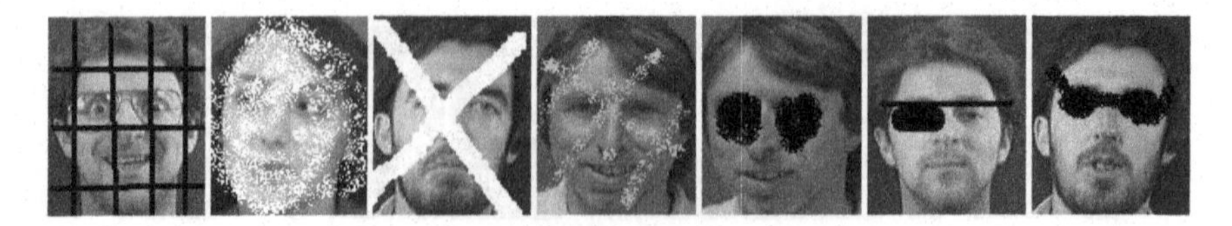

Fig. 7.16 Occluded images which are only recognized by *CICA*-based recognition system, but can not be recognized by ***R****ICA* and any *PCA*-based system

Fig. 7.17 Blurred images which are not recognized with any feature extraction technique

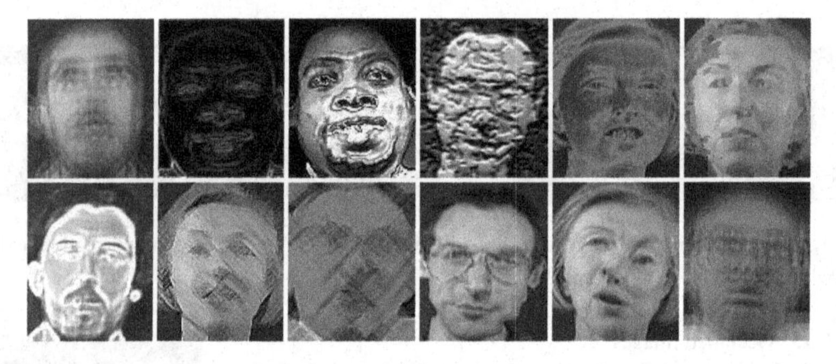

Fig. 7.18 Blurred images which are only recognized by *CICA*-based recognition system, but can not be recognized by ***R****I C A* and any *PCA*-based system

Fig. 7.19 Blurred images which are not recognized by any *PCA*-based recognition system

observed that only *CICA*-based recognition system can correctly classify these faces while other techniques including ***R****ICA* failed to do so. Further Fig. 7.19 presents the set of images possessing lesser blurring in comparison to Fig. 7.18 are again not recognized by any *PCA*-based recognition system but they are well recognized by ***R****ICA* and *CICA*-based recognizer. It turns out the fact from these observations that ***R****ICA* is better than both PCA but *CICA* is best among all. Thus, *CICA* provides more discriminating power among the classes than any other technique, it is due to robust class distinctiveness yielded by *CICA*-based recognition system.

As with any other classification studies, the class distinctiveness must be taken into account for comparing different feature extraction techniques. In order to further access the robustness of different feature extraction techniques to noise, distortion, and other environmental effects, I carried out a comparative assessment of different techniques over electronically modified [14] coloured images. Figure 7.20 presents a set of images in which low distortion is introduced. These images are passed to previously trained ***C*RSP** classifiers with different feature extraction techniques. ***CICA*** and ***RICA***-based classifiers are able to correctly classify them while any PCA-based classifiers are not able to recognize them. Another set of images in Fig. 7.21 have comparatively more distortions and variations; these images can not be recognized by any PCA and ***RICA***-based recognizer, while *CICA*-based recognition system is able to correctly identify them. It is due to better class distinctiveness yielded by ***CICA***. Though, there are limitations in *CICA*-based technique as there are always in every methods. Figure 7.22 presents some images with very high degree of distortions, where *CICA*-based system also failed to perform proper recognition. Therefore, the basis vectors obtained by ***CICA*** is superior to ***RICA*** and PCA in the sense that it provides a representation (feature vector), which is more robust to the effect of noise. It is therefore observed in all experiments that *CICA* outperforms over other popular techniques, PCA and ***RICA***, for recognition specially in noisy environment.

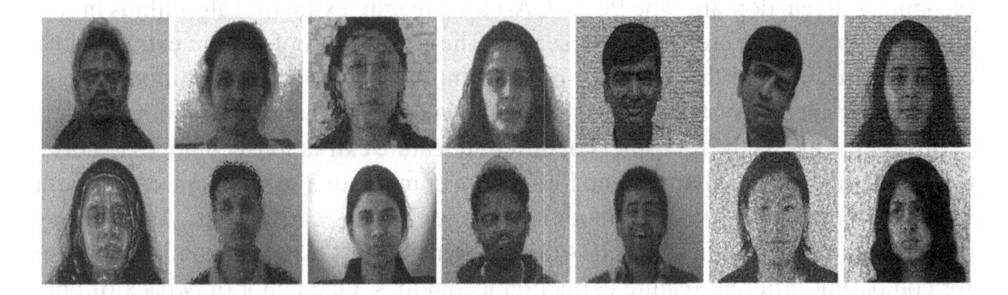

Fig. 7.20 Distorted images which are not recognized with any *PCA*-based recognition system. But, recognized by ***RICA*** and ***CICA***-based recognition system

Fig. 7.21 Distorted images which are recognized with only *CICA*-based recognizer

Fig. 7.22 Distorted images which are not recognized with any feature extraction-based technique presented in this book

7.7 Inferences and Discussion

This chapter is devoted for designing machine learning techniques in complex domain. The most important contributions are presenting a new class of classifier "OCON," which contains the ensemble of higher-order neurons in complex domain and a new feature extractor "***C***ICA" which yielded surprisingly good class distinctiveness among different subjects in image database. The remarkable achievements in the proposed recognizer (classifier) is its compact structure, improved learning speed, lesser weights storage and better accuracy in recognition, which contains the ensemble of single ***C***RSS and ***C***RSP neurons instead of ensemble of neural networks ie "OCONN." It is worth mentioning that its overall performance is much better than recognizer in real domain. The PCA-ICA-based feature extraction algorithms in real and complex domain are further presented. The capabilities of feature extractor and classifier are justified through their assessment in typical face recognition system. The performance of system is tabulated with major performance evaluation metrics.

In order to assess the robustness of different feature extraction techniques to noise, distortion, and other environmental effects, a comparative assessment of different techniques over electronically modified images [14] is carried out. As with any other classification studies, the class distinctiveness must also be taken into account for comparing different feature extraction techniques. One major drawback of both PCA-based feature extraction methods is that distinctiveness among classes during classification is very poor. They are not able to classify the occluded and blurred face images considered in experiments. This shows that real and complex PCA are not suitable for design of recognition system which can work in real environmental applications. Therefore, the basis vectors obtained by ***C****ICA* is superior to ***RICA*** and PCA in the sense that it provides a representation (feature vector), which yields far better class distinctiveness, hence is more robust to the effect of noise. It is therefore observed in this chapter that ***C****ICA* outperforms over other techniques for recognition especially in noisy environment.

References

1. Daugman, J.: Face dection : a survey. Comput. Vis. Image Underst. **83**(3), 236–274
2. Chellappa, R., Wilson, C.L., Sirohey, S.: Human and machine recognition of faces: a survey. Proc. IEEE **83**, 705–740 (1995)
3. Kong, S.G., Heo, J., Abidi, B.R., Paik, J., Abidi, M.A.: Recent advances in visual and infrared face recognition-a review. Comput. Vis. Image Underst. **97**, 103–135 (2005)
4. Bhattacharjee, D., Basu, D.K., Nasipuri, N., Kundu, M.: Human face recognition using multilayer perceptron. Soft Comput. **14**, 559–570 (2010)
5. Abate, A.F., Nappi, M., Riccio, D., Sabatino, G.: 2D and 3D face recognition: a survey. Pattern Recogn. Lett. **28**, 1885–1906 (2007)
6. Chen, L.F., Liao, H.M., Lin, J., Han, C.: Why recognition in a statistic-based face recognition system should be based on the pure face portion: a probabilistic decision-based proof. Pattern Recogn. **34**(7), 1393–1403 (2001)
7. Sebe, N., Lew, M.S., Huijsmans, D.P.: Toward improved ranking metrics. IEEE Trans. Pattern Anal. Mach. Intell. **22**(10), 1132–1143 (2000)
8. Rattan, S.S.P., Hsieh, W.W.: Complex-valued neural networks for nonlinear complex principal component analysis. Neural Networks **18**, 61–96 (2005)
9. Anemuller, J., Sejnowski, T., Makeig, S.: Complex independent component analysis of frequency-domain electroencephalographic data. Neural Networks **16**(9), 1311–1323 (2003)
10. Calhoun, V.D., Adal, T.: Unmixing fMRI with independent component analysis. IEEE Eng. Med. Biol. Mag. **25**(2), 79–90 (2006)
11. Hietmeyer, R.: Biometric identification promises fast and secure processing of airline passengers. Int. Civil Aviat. Org. J. **55**(9), 10–11 (2000)
12. Hahn, S.L.: Hilbert Transforms in Signal Processing. Artech House, Boston, MA (1996)
13. Lanitis, A., Taylor, C.J., Cootes, T.F.: Towards automatic simulation of ageing effects on face images. IEEE Trans. Pattern Anal. Mach. Intell. **24**(4), 442–455 (2002)
14. Zhao, W., Chellappa, R., Phillips, P.J., Rosenfeld, A.: Face recognition: a literature Survey. ACM Comput. Surv. **35**(4), 399–458 (2003)
15. Wiskott, L., Fellous, J.M., Kruger, K., Von der Malsburg, C.: Face recognition by elastic bunch graph matching. IEEE Trans. Pattern Anal. Mach. Intell. **19**(7), 775–779 (1997)
16. Vetter, T., Poggio, T.: Face recognition by elastic bunch graph matching. IEEE Trans. Pattern Anal. Mach. Intell. **19**(7), 733–742 (1997)
17. Ara, V.N., Monson, H.H.: Face recognition using an embedded HMM. In: Proceedings of International Conference on Audio- and Video-Based Biometric Person Authentication, pp. 19–24 (1999)
18. Bevilacqua, V., Cariello, L., Carro, G., Daleno, D., Mastronardi, G.: A face recognition system based on Pseudo 2D HMM applied to neural network coefficients. Soft Comput. **12**, 615–621 (2008)
19. Volker, B., Sami, R., Thomas, V.: Face identification across different poses and illuminations with a 3D morphable model. In: Proceedings of IEEE International Conference on Automatic Face and Gesture Recognition, pp. 202–207 (2002)
20. Cover, T., Thomas, J.: Elements of Information Theory. John Wiley and Sons, Chichester (1991)
21. Marcialis, G.L., Roli, F.: Fusion of appearance based face recognition algorithms. Pattern Anal. Appl. **7**(2), 151–163 (2004)
22. Cevikalp, H., Neamtu, M., Wilkes, M., Barkana, A.: Discriminative common vectors for face recognition. IEEE Trans. PAMI **27**(1), 1–9 (2005)
23. Kirby, M., Sirovich, L.: Application of the Karhunen-Loeve procedure for the characterization of human faces. IEEE Trans. Pattern Anal. Mach. Intell. **12**(1), 103–108 (1990)
24. Turk, M., Pentland, A.: Eigenfaces for recognition. J. Cogn. Neurosci. **3**(1), 71–86 (1991)
25. Savvides, M, Vijaya Kumar, B.V.K., Khosla, P.K.: Eigenphases vs. eigenfaces. In: Proceedings of 17th International Conference on Pattern Recognition, vol. 3, pp. 810–813 (2004)

26. Ahonen, T,. Pietikainen, M., Hadid, A.: Face recognition based on the appearance of local regions. In: Proceedings of 17th International Conference on Pattern Recognition, vol. 3, pp. 153–156 (2004)
27. Liu, C.: Gabor-based kernel PCA with fractional power polynomial models for face recognition. IEEE Trans. Pattern Anal. Mach. Intell. **26**(5), 572–581 (2004)
28. Martinez, A.M., Kak, A.C.: PCA versus LDA. IEEE Trans. Pattern Anal. Mach. Intell. **23**(2), 228–233 (2001)
29. Kim, T., Kittler, J.: Locally linear discriminant analysis for multimodally distributed classes for face recognition with a single model image. IEEE Trans. Pattern Anal. Mach. Intell. **27**(3), 318–327 (2005)
30. Dai, G., Qian, Y., Jia, S.: A Kernel fractional-step nonlinear discriminant analysis for pattern recognition. In: Proceedings of 17th International Conference on Pattern Recognition, vol. 2, pp. 431–434 (2004)
31. Bartlett, M.S., Lades, H.M., Sejnowski, T.J.: Independent component representations for face recognition. Proc. of SPIE **3299**, 528–539 (1998)
32. Cichocki, A., Amari, S.: Adaptive Blind Signal and Image Processing: Learning Algorithms and Applications. John Wiley and Sons, New York (2002)
33. Kwak, K., Pedrycz, W.: Face recognition using an enhanced independent component analysis approach. IEEE Trans. Neural Network **18**(2), 530–541 (2007)
34. Kim, J., Choi, J., Yi, J., Turk, M.: Effective representation using ICA for face recognition robust to local distortion and partial occlusion. IEEE Trans. Pattern Anal. Mach. Intell. **27**(12), 1977–1981 (2005)
35. Hyvarinen, A., Karhunen, J., Oja, E.: Independent Component Analysis. John Wiley and Sons, New York (2001)
36. Huang, D., Mi, J.: A new constrained independent component analysis method. IEEE Trans. Neural Netw. **18**(5), 1532–1535 (2007)
37. Wei, L., Jagath, C.R.: ICA with reference. Neurocomputing **69**, 2244–2257 (2006)
38. Bartlett, M.S., Movellan, J.R., Sejnowski, T.J.: Face recognition by independent component analysis. IEEE Trans. Neural Netw. **13**(6), 1450–1464 (2002)
39. Er, M.J., Chen, W., Wu, S.: High-speed face recognition based on discrete cosine transform and RBF neural networks. IEEE Trans. Neural Netw. **16**(3), 679–691 (2005)
40. Hyvarinen, A.: Survey on independent component analysis. Neural Comput. Surv. **2**, 94–128 (1999)
41. Bell, A.J., Sejnowski, T.J.: An information-maximization approach to blind separation and blind deconvolution. Neural Comput. **7**(6), 1129–1159 (1995)
42. Horel, J.D.: Complex principal component analysis: theory and examples. J. Clim. Appl. Meteorol. **23**, 1660–1673 (1984)
43. Jan, E.M., Visa, K.: Complex random vectors and ICA models: identifiability, uniqueness and separability. IEEE Trans. Inf. Theory **52**(3), 596–609 (2006)
44. Adal, T., Kim, T., Calhoun, V.: Independent component analysis by complex nonlinearities. In: Proceedings of International Conference on Acoustics Speech Signal Process, Montreal, ON, Canada, May 2004, vol. 5, pp. 525–528 (2004)
45. Huang, N.E.: Introduction to Hilbert-Huang transform and its associated mathematical problems. In: Huang, N.E., Attoh-Okine, N. (eds.) Hilbert-Huang Transform in Engineering, pp. 1–32. CRC Press, New York (2005)
46. Amores, J., Sebe, N., Radeva, D.P.: Boosting the distance estimation: application to the K-nearest neighbor classifier. Pattern Recogn. Lett. **27**, 201 (2006)
47. Lawrence, S., Lee, G.C., Ah, C.T., Andrew, D.B.: Face recognition: a convolutional neural network approach. IEEE Trans. Neural Netw. **8**, 98–113 (1997). (special issue on neural networks and pattern recognition)
48. Oh, B.J.: Face recognition by using neural network classifiers based on PCA and LDA. In: Proceedings of IEEE International Conference on Systems, Man and Cybernetics, pp. 1699–1703, 10–12 Oct 2005

49. Sing, J.K., Basu, D.K., Nasipuri, M., Kundu, M.: Face recognition using point symmetry distance-based RBF network. Appl. Soft Comput. **7**, 58–70 (2007)
50. Bishop, C.M.: Neural Networks for Pattern Recognition. Clarendon Press, Oxford (1995)
51. Aitkenheada, M.J., Mcdonald, A.J.S.: A neural network face recognition system. Eng. Appl. Artif. Intell. **16**(3), 167–176 (2003)
52. Giacinto, G., Roli, F., Fumera, G.: Unsupervised learning of neural network ensembles for image classification. IEEE IJCNN **3**, 155–159 (2000)
53. Haddadnia, J., Faez, K., Moallem, P.: Neural network based face recognition with moments invariant. In: Proceedings of IEEE International Conference on Image Processing, vol. I, pp. 1018–1021, Thessaloniki, Greece, 7–10 Oct 2001
54. Lee, T.W.: Independent Component Analysis: Theory and Applications. Kluwer, Boston, MA (1998)
55. Oja, E., Yuan, Z.: The FastICA algorithm revisited: convergence analysis. IEEE Trans. Neural Networks **17**(6), 1370–1381 (2006)
56. Nitta, T.: An extension of the back-propagation algorithm to complex numbers. Neural Netw. **10**(8), 1391–1415 (1997)
57. Calhoun, V., Adali, T.: Complex infomax: convergence and approximation of infomax with complex nonlinearities. VLSI Signal Process. Springer Sci. **44**, 173–190 (2006)
58. Calhoun, V., Adali, T., Pearlson, G.D., Pekar, J.J.: ON complex infomax applied to complex FMRI data. In: Proceedings of ICASSP, Orlando, FL (2002)
59. Brown, J.W., Churchill, R.V.: Complex Variables and Applications, 7th edn. Mc Graw Hill, New York (2003)
60. Saff, E.B., Snider.: Fundamentals of Complex Analysis with Applications to Engineering and Science. Englewood Cliffs, New Jersey (2003)
61. Novey, M., Tlay, A.: Complex ICA by negentropy maximization. IEEE Trans. Neural Network **19**(4), 596–609 (2008)
62. ORL face database.: http://www.uk.research.att.com/facedatabase.html
63. Vidit, J., Amitabha, M.: The Indian face database. http://vis-www.cs.umass.edu/vidit/IndianFaceDatabase (2002)

Printed in the United States
By Bookmasters